SpringerBriefs in Business

SpringerBriefs present concise summaries of cutting-edge research and practical applications across a wide spectrum of fields. Featuring compact volumes of 50 to 125 pages, the series covers a range of content from professional to academic. Typical topics might include:

- A timely report of state-of-the art analytical techniques
- A bridge between new research results, as published in journal articles, and a contextual literature review
- A snapshot of a hot or emerging topic
- An in-depth case study or clinical example
- A presentation of core concepts that students must understand in order to make independent contributions

SpringerBriefs in Business showcase emerging theory, empirical research, and practical application in management, finance, entrepreneurship, marketing, operations research, and related fields, from a global author community.

Briefs are characterized by fast, global electronic dissemination, standard publishing contracts, standardized manuscript preparation and formatting guidelines, and expedited production schedules.

Marival Segarra-Oña • Virginia Santamarina-Campos • Ángel Peiró-Signes
Editors

Managing the Transition to a Circular Economy

Action Plans in the Tourism Sector

Editors
Marival Segarra-Oña
Department of Business Organisation
Universitat Politècnica de València
Valencia, Spain

Virginia Santamarina-Campos
Department of Conservation and Restoration of Cultural Assets
Universitat Politècnica de València
Valencia, Spain

Ángel Peiró-Signes
Department of Business Organisation
Universitat Politècnica de València
Valencia, Spain

ISSN 2191-5482 ISSN 2191-5490 (electronic)
SpringerBriefs in Business
ISBN 978-3-031-49688-2 ISBN 978-3-031-49689-9 (eBook)
https://doi.org/10.1007/978-3-031-49689-9

© The Editor(s) (if applicable) and The Author(s) 2024. This book is an open access publication.
Open Access This book is licensed under the terms of the Creative Commons Attribution 4.0 International License (http://creativecommons.org/licenses/by/4.0/), which permits use, sharing, adaptation, distribution and reproduction in any medium or format, as long as you give appropriate credit to the original author(s) and the source, provide a link to the Creative Commons license and indicate if changes were made.

The images or other third party material in this book are included in the book's Creative Commons license, unless indicated otherwise in a credit line to the material. If material is not included in the book's Creative Commons license and your intended use is not permitted by statutory regulation or exceeds the permitted use, you will need to obtain permission directly from the copyright holder.

The use of general descriptive names, registered names, trademarks, service marks, etc. in this publication does not imply, even in the absence of a specific statement, that such names are exempt from the relevant protective laws and regulations and therefore free for general use.

The publisher, the authors, and the editors are safe to assume that the advice and information in this book are believed to be true and accurate at the date of publication. Neither the publisher nor the authors or the editors give a warranty, expressed or implied, with respect to the material contained herein or for any errors or omissions that may have been made. The publisher remains neutral with regard to jurisdictional claims in published maps and institutional affiliations.

This Springer imprint is published by the registered company Springer Nature Switzerland AG
The registered company address is: Gewerbestrasse 11, 6330 Cham, Switzerland

Paper in this product is recyclable

Foreword

Towards a Tourism Model Framed Within the Circular Economy

The apocalyptic narrative of the new prophets of doom need not be taken at face value, and much less should all their terrifying utterances be accepted. However, any reasonably well-informed citizen can see for him or herself the undeniable reality that humanity is facing a worrying scenario, to say the least.

Leaving aside the war in Ukraine and all the other geopolitical tensions in the world, the fact is that we are immersed in a widespread, polymorphic crisis, with a heterogeneous multitude of economic, energy, cultural, ethical and axiological derivatives. In fact, given the daunting scale of the challenges posed by each of these issues, it would seem that our way of life and civilisation as a whole should be called into question.

Perhaps the most alarming version of this multifaceted crisis is the formidable environmental challenge facing humanity: a challenge associated with the serious risk of ending up causing irreversible and irreparable damage to ecological balances on which depend not only the viability of natural systems but even the continuity of life itself on the planet as we know it.

In fact, in academic discussions it has become little more than commonplace, albeit controversial, to debate whether humanity has not already been living for decades in a new historical epoch or even in an incipient geological era, the Anthropocene, the result not so much of telluric activity or physical-cosmic forces, but above all as an unintended by-product of the clearly unsustainable impact of reckless human activity on the environment. Moreover, there is no prospect of a solution to the problem, at least in the short term, unless measures are taken to alleviate some of the more dramatic epiphenomena of this kind of mad rush towards the abyss. These include, for example, air and water pollution, the increase in waste and litter, the depletion of non-renewable resources, the loss of biodiversity and the fact that climate change could lead to the collapse of the ecosystem.

Beyond ideological controversies and sterile discussions, often lacking in the necessary rationality and dispassion, the fact is that, on the basis of objective data that are increasingly difficult to question, a state of opinion is consolidating on a global scale that is increasingly demanding the collaboration of the agents involved—public administrations, civil society institutions and, of course, companies and economic organisations—in the common, urgent task of tackling the ecological problem in a decisive and rigorous manner. At all events, rather than looking for culprits, it is a matter of finding innovative solutions based on the resources available and the synergies that can be expected from creativity and collaboration between the various stakeholders.

When one studies economic history, one becomes aware of a kind of economic miracle of humankind which, starting with the First Industrial Revolution, made it possible to overcome the more than pitiful levels of destitution and hardship in which large sections of the population had lived until then. Unfortunately, millions of people—old people, adults and children—are still starving all around the world today. And many more are still living miserably below what we call the poverty line.

Leaving aside other considerations, the solution to such dramatic circumstances seems to require economic growth as a precondition. Only through the efficient production of goods would it seem possible to satisfy human needs, both the most basic and the more sophisticated. However, at this point we come up against a double paradox that should be made explicit.

Firstly, while it is true that industrialisation and the modern factory favoured economic growth which, in turn, made effective human development possible, they did so by leveraging what is known as the carbon economy. Let us remember that the new energy source in the mid-eighteenth century, steam, required the burning of coal. However, given the times we live in—the Fourth Industrial Revolution—it does not seem sustainable or desirable to try to follow the path that, at least in the West, we have been following for the last 250 years. And this, as I have said, without prejudice to recognising that the process of economic growth and social development has been successful until the last quarter of the last century, at least for the industrialised countries, when anomalies in the model began to be recognised.

The second dimension of the paradox has to do with another complementary aspect: the way in which the productive process is carried out and which constitutes the usual practice in the manufacture of most of the goods that end up on the market. The *modus operandi* of this "linear economy" is ordered according to a process whose logic offers a reasonable sequence of moments and tasks that can be efficiently managed.

To put it very simply and without going into specifics, the life of a given consumer good, whether it is a pencil, a hybrid car or a sophisticated next-generation smartwatch, could be described accurately by the following linear sequence: (1) extraction of the raw materials from which to make the final product; (2) transport along supply chains to production plants; (3) manufacture of the products; (4) distribution of the goods for sale; (5) purchase and use by the consumer or end-user; (6) elimination and disposal of the residues—turned into waste and unusable material—once the product has outlived its useful lifetime.

This sequence of production-distribution-consumption-waste is the norm in all economic contexts and would probably not even have been called into question if three complementary circumstances had not come together: firstly, the emergence of sensitivity to the ecological problem, as evidenced by egregious pollution of the environment and our awareness of the limits of the planet's resources. Secondly, the technical unsustainability of economic growth guided by the same parameters. And above all, thirdly, if, in the face of foreseeable—and undesirable—social consequences, the practices associated with this technical form of production and the prevailing culture of consumption of what is produced and disposing of waste had not previously been questioned from an ethical standpoint.

All this is leading to a thorough reconsideration of the logic of the economic system and, naturally, to a search for alternative or, at least, more intelligent ways of moving towards what has come to be known as sustainable development. That type of development which, in the 1980s, was already defined as "development that meets the needs of the present generation without jeopardising the needs of future generations".

An excessive focus on production, the inevitable counterpart of excessive consumerism, in the wake of the throwaway culture, is now being seen as unsustainable from a technical point of view and, gradually, more problematic from an ethical point of view. This new and increasingly common view is perceptible at least in developed countries and among certain groups who are unhappy that, for the sake of short-term and selfish economic gain, the wheel of economic growth continues to spin faster. This is in fact exacerbating the depletion of resources and further contributing to the deterioration of the planet, to the serious detriment of humanity as a whole.

In response to the deteriorating ecological problem resulting from the dual framework within which most companies operate, that is, the Linear Economy and the Carbon Economy, alternative narratives to business as usual are beginning to be constructed as a contribution to the redesign and improvement of the economic dynamic, in a bid to make it fairer and more sustainable. Initiatives promoted by different bodies and institutions, including governments, public administrations and supranational bodies such as the European Union, the UN and the OECD, seem to converge in this objective. In addition to NGOs, there are also many companies and economic institutions—investment funds, regulatory bodies, etc.—and the business schools themselves.

Indeed, it is evident how the business narrative has been evolving towards a concept of management that is more in line with the new realities and hence the insistence on identifying and maintaining organisational purpose as a reference point from which to design a long-term strategy. The new cultural climate in which economic activity is taking place is also shedding light on efforts to propose an expanded mission for the company and, consequently, also for the professional activity of entrepreneurs, managers and members of boards of directors. This broadening of objectives entails accepting that company management should primarily not seek only the maximisation of economic profit in the short term, to the benefit of the owners and shareholders.

Therefore, without neglecting the financial objective, for the company and its management to truly contribute to the common good, bringing closer the goals identified as tangible manifestations of the rhetoric of the Sustainable Development Goals (SDG), that financial objective should be framed in a broader perspective. Firstly, in the vision that aims to optimise the Triple Bottom Line, not only in terms of the results expected from the successful management of economic capital, but also in terms of the objectives that would be expected from good management of the other two types of capital that the company uses in its activity: social capital and environmental capital.

In line with this is the legislation requiring reporting, that is, legislation that obliges companies to report to society in an integrated, holistic, transparent and comprehensible manner on their positive and negative impacts in the three dimensions mentioned above. The ESG criteria—environmental, social and governance—when designing business strategies are also being considered in the light of the above-mentioned expectations.

For several decades, business theory in academic circles and in the business management model has also been evolving in practice towards an interest in serving the broader base of the company. Indeed, instead of operating exclusively for the economic and immediate interests of shareholders, it also aspires to attend to the legitimate interests of the other stakeholders (including customers, workers, suppliers, distributors...) who, moreover, make the organisational dynamics possible and from whose satisfaction derive both economic success and a licence to operate, a good reputation and, ultimately, the maintenance of the company in a global market and within the framework of an efficient, equitable, sustainable economy. This would not be possible without a strategy that, in some way, connects with a moral choice to respect the environment and care for creation.

Therefore, it is within this framework that the current appeal to the need to move from a linear economy to an alternative model is to be found, a model whose theme is the powerful metaphor represented by the Circular Economy label. In fact, it is beginning to make headway as an innovative, creative project offering the opportunity to design and implement new business models which should eventually lead to an economic paradigm that is in keeping with the times: more efficient and fairer but also, above all, truly sustainable.

Fortunately, the legislative initiatives promoted by the European Union are already making some progress in this direction, albeit with evident parsimony and with a certain lack of coordination in the action plans. In fact, in the case of Spain, despite the existence of the *Spanish Strategy for a Circular Economy 2030,* there is still a long way to go, both in the harmonisation of the legislation of the different regional governments and, above all, in a kind of pedagogical task that, through the training of businessmen and entrepreneurs, on the one hand, and raising public awareness, on the other, may eventually lead to a real cultural change.

Respect for the environment should be assumed as a condition of possibility for true social progress, capable of contributing to real human development. For the moment, it seems that the discourse is being articulated by reference to the watchword of sustainability. What must be done is to take control of the narrative and move it on from the superficial tone that characterises it at the moment and get to the heart of the matter. It would be advisable to help to move this narrative forward, beyond the marketing and cosmetics of hyperbolic, overblown advertisements, in which everything, almost without exception, is presented as "sustainable". If this trivialisation of the concept is not overcome, the risk is threefold: firstly, to empty it of content; secondly, to generate in the receiver of the message acute scepticism with regard, if not to the seriousness of the problem, then to the seriousness with which economic agents are willing to respond to the challenge. And furthermore, thirdly, there is the risk of having missed the opportunity to redesign the economic model to make it sustainable and contribute to the evolution of a cultural model in which people and societies can find more favourable opportunities to flourish.

Fortunately, there are also many other proposals under way in addition to the current superficial, insubstantial initiatives, especially in the primary sector and, above all, in the economy of consumer goods production. It is obvious that in the latter, more than in the services sector, it is easier and more practical to redesign products and processes to enable reuse, repair and recycling. In fact, natural dynamics itself offers clues to progress, from the moment one realises that nature never produces waste. Everything is interconnected and integrated into a mega-cycle in which, for example, waste of any kind never ends up as waste, but is immediately converted into new resources through which the necessary ecological balance is maintained, making the maintenance of life itself possible. From this empirical evidence, it is possible to establish a kind of metaphysical axiom, from which the theoretical assertion of entropy and the Second Principle of Thermodynamics takes on its full practical potential.

The deep structure of the question that should kindle entrepreneurs' creativity in the Circular Economy could be delineated by reference to a reasoning in which, as premises, the following four would stand out as the starting points: (1) in nature, energy is neither created nor destroyed, but simply transformed, thereby maintaining the ecological balance; (2) on the other hand, the linear economic model irreparably destroys resources that are not only scarce but, above all, limited, and for which there is no known replacement or renewal, not even (at least for the time being) with the help of the digitalisation of economic activity; (3) planetary ecology is beginning to become unbalanced, having suffered from this unsustainable economic activity; (4) the viability of human life continues to require that needs be satisfied through economic activity and efficient management for environmentally friendly production, the efficient, fair distribution of what is produced, responsible consumption and special attention to the way in which waste can be converted into new resources.

Assuming the above, the challenge of innovation has been thrown down and, at the same time, so has the stimulus to develop new ways of thinking that can give rise to heterodox approaches or, at least, approaches that are different from the mainstream of normal science in economics and business as usual in business praxis.

Customers, investors and competitors themselves are turning towards sustainable models of economics and management, and the Circular Economy is beginning to demand its rights, while offering opportunities to those who demonstrate a broad vision of business and management.

Nowadays, start-ups of all kinds—especially those that are being developed on the basis of entrepreneurial initiatives in the context of the digitalisation of the economy—are already tending to this orientation towards circularity. Consequently, making a virtue out of necessity and without waiting for legislation to force them to adopt certain types of practices, the most innovative companies are seriously asking themselves profound questions, the answers to which will undoubtedly lead to business models that will enable them to lead the market in the medium term. This is precisely because they have taken the ecological challenge seriously and have found a way to turn some of the opportunities that arise when care for the planet becomes an organisational purpose into a business.

When deciding to take the path of the Circular Economy, the key issue needs to be addressed by coming up with a reasonably convincing answer to a first question. In this sense, it is possible, without seeing them as mutually exclusive, to adduce reasons of economic sustainability, of a moral willingness to respect nature, as a creative response to the challenge of innovation to structure an economic activity that neither consumes resources nor generates waste. However, once the first *why* has been sufficiently answered, the more operational question of *how* immediately arises.

Fortunately, anyone wishing to join in the design and implementation of sustainable business models and, more specifically, based on the theoretical postulates of the Circular Economy, does not have to start from scratch. In fact, there are already many examples of benchmarks and good practices in which to seek inspiration. In this regard, it could be said that perhaps one of the most opportune contributions that academia could make in the present circumstances with respect to the common task of organising a better response to the Economic Imperative—that is, a more efficient, equitable, responsible and sustainable response—could be precisely that of researching with methodological solvency, developing theoretical models and, above all, transferring the results of the research and disseminating the knowledge that allows economic agents, administrative authorities and citizens to find their bearings in these fascinating domains.

The book to which these pages serve as a prologue addresses the issue of how the transition from a linear economy to another model is being implemented, based on the postulates and principles of the Circular Economy. It is the result of the line of research carried out within the framework of the *InnoEcotur* project. *Circular Economy Strategy in the Tourism Sector of the Valencian Community* https:// innoecotur.webs.upv.es/.

Readers interested in the practical aspects will find in the chapters of this book a selection of initiatives and good business practices in the tourism sector of a thriving and innovative Spanish region. Beyond this, the book, which is in itself a good inspirational prototype, will also stimulate theoretical analysis on the basis of future work and lines of research to be extrapolated to other regional contexts by other groups, both in Spanish universities and in the wider international context.

Iberdrola Chair in Economic and
Business Ethics, Faculty of Economics
and Business Administration—ICADE,
Pontifical University Comillas, Madrid,
Spain

José Luis Fernández Fernández

Acknowledgments

This book has been supported by the Regional Ministry of Innovation, Universities, Science and Digital Society of the Valencian Regional Government (CIAORG/2021/39).

Contents

1 **Introduction: Towards a Circular Economy Strategy in the Valencian Region's Tourism Industry**.................... 1
Marival Segarra-Oña and Virginia Santamarina-Campos

Part I Challenges and Opportunities

2 **The Transition Journey to the Circular Economy by Companies in the Valencian Region's Tourism Industry**...... 7
Blanca de-Miguel-Molina, María de-Miguel-Molina, Luis Miret-Pastor, and María Belén Silva-Cárdenas

3 **Impact Culture and the Circular Economy in the Tourism Industry: An Analysis of Challenges and Recommendations for Sustainability**.. 19
Virginia Santamarina-Campos, Miguel Ángel Mas-Gil, María de-Miguel-Molina, and Daniel Catalá-Pérez

4 **Designing a Dynamic Map of Circular Economy in the Tourism Sector of the Valencian Community**...................... 33
Conrado Carrascosa-Lopez, M. Rosario Perello-Marin, and María Ángeles Carabal-Montagud

Part II Good Practices

5 **Development of a Model for the Application of the Circular Economy in Hotels and Restaurants Through the 'Customer Journey Map'**.. 47
Joaquín Sánchez-Planelles, Yolanda Trujillo-Adriá, and Gabriela Ribes-Giner

6 Wine Tourism, Circular Economy Practices and Hospitality
   in the Spanish Wine Industry: The Case of Bodegas Casa Sicilia
   Wine Restaurant...................................... 61
   Bartolomé Marco-Lajara, Javier Martínez-Falcó,
   Eduardo Sánchez-García, and Luis A. Millán-Tudela

7 Circular Economy Practices in the Spanish Beer Industry:
   The Case of the Beer Producer La Somniada................. 69
   Francisco Puig, Guillermo Navarro-Sanfelix, and Santiago Cantarero

8 Good Practices of Circular Economy in Tourism in Castellón..... 79
   Andrei Serbanescu, Luís Martínez Cháfer,
   and Teresa Martínez Fernández

Part III Research, Innovation, Competitiveness and Production

9 Importance of Culture and Innovation in Behaviors
   Towards the Circular Economy in Spanish Hotels............. 91
   Bartolomé Marco-Lajara, Mercedes Úbeda-García,
   Esther Poveda-Pareja, and Encarnación Manresa-Marhuenda

10 Circular Economy Self-assessment Tool for Hotels............. 101
   Marival Segarra-Oña, Ángel Peiró-Signes, Joaquín Sánchez-Planelles,
   and Esther Poveda-Pareja

11 Conclusions: Tourism Sustainability and Improvement Plans..... 119
   Ángel Peiro-Signes and Virginia Santamarina-Campos

Abbreviations

ADR	Average Daily Rate
AVE	Average Variance Extracted
AVEN	Valencian Energy Agency
C	Category
CDCA	Documentation Center for Environmental Conflicts
CE	Circular Economy
CEP	Circular Economy Practices
CEPs	Corporate Environmental Practices
CI	Confidence Interval
CO_2	Carbon Dioxide
e.g.	for example
EC	European Commission
ECG	Economy for the Common Good
ESG	Environmental, Social and Governance
EU	European Union
GDP	Gross Domestic Product
GHG	Green House Gas
GIEC	Interplatform Group for a Circular Economy
GINN	Green Innovation
H1	Hypothesis 1
H2	Hypothesis 2
HTMT	Heterotrait-Monotrait Ratio
i.e.	that is
ICT	Information and Communication Technologies
OC	Organisational Culture
Occ	Occupancy
P	Proposition
PAR	Participatory Action Research
PERF	Corporate Performance
PLS	Partial Least Squares

PLS-SEM	Partial Least Squares Structural Equation Modelling
Q	Question
R&D	Research and Development
RevPAR	Revenue Per Available Room
RQ	Research Question
RQs	Research Questions
Rs	Redesign, Reduce, Reuse, Renovate/Repair, Restore/Remanufacture, Recover/Return and Recycle
RSC	Corporate Social Responsibility
SABI	System for Analysis of Iberian Balances
SDG	Sustainable Development Goals
SDG11	Sustainable cities and communities
SDG12	Responsible consumption and production
SDG13	Climate action
SDG14	Life below water
SDG15	Life and land
SDG6	Clean water and sanitation
SDG7	Affordable and clean energy
SDGs	Sustainable Development Goals
SEM	Structural Equation Modelling
SMEs	Small and Medium-sized Enterprises
SP	Sustainable Performance
STDV	Standard Deviation
UA	University of Alicante
UJI	Universitat Jaume I
UN	United Nations
UNWTO	United Nations World Tourism Organization
UPV	Universitat Politècnica de València
UV	Universitat de València
VIF	Variance Inflation Factor
WT	Wine Tourism
WTTC	World Travel & Tourism Council

Chapter 1
Introduction: Towards a Circular Economy Strategy in the Valencian Region's Tourism Industry

Marival Segarra-Oña and Virginia Santamarina-Campos

Although the tourism industry is a cornerstone of the Valencian Region's economy, it generates a variety of environmental impacts that require robust management and the implementation of measures to mitigate them. To address this issue, an innovative project entitled "Creation of an Innovation Platform for the Promotion and Implementation of a Circular Economy Strategy in the Valencian Region's Tourism Industry" has been devised.

This ambitious project, called InnoEcoTur, has received funding from the Valencian Innovation Agency and is being rolled out under the expert coordination of the Universitat Politècnica de València (UPV). The project has also benefited from significant contributions made by members of the University of Alicante (UA), the Universitat Jaume I (UJI) and the Universitat de València (UV).

The main objective of InnoEcoTur is the creation of an Innovation Platform for the circular economy in the Valencian Region to promote circular economy models in the tourism industry. To achieve this objective, the project seeks to involve different stakeholders and focus on the development of sectoral indicators, methodologies and tools that enable companies to comprehensively assess the impact of their activities from an environmental footprint perspective.

In addition, InnoEcoTur aims to create jobs in the circular economy and provide support and training to specialists to develop new environmental consultancy services. These efforts culminate in the fundamental objective of the project: to promote the circular economy in the tourism industry through the transfer of results and the implementation of eco-innovations and circular economy initiatives across the

---

M. Segarra-Oña
Department of Business Organisation, Universitat Politècnica de València, Valencia, Spain
e-mail: maseo@omp.upv.es

V. Santamarina-Campos (✉)
Department of Conservation and Restoration of Cultural Assets, Universitat Politècnica de València, Valencia, Spain
e-mail: virsanca@upv.es

board, ranging from production and processes to business models and organizational practices.

InnoEcoTur is working closely with key stakeholders in the Valencian Region's tourism industry. This project has several specific objectives, including the creation of a dynamic map of the circular economy in the tourism sector, drawing up a catalogue of good practices with eco-innovative technologies and results, and the creation of a collaborative platform to encourage cooperation between the various stakeholders in the Valencian Region.

Two main strategies are being pursued to facilitate the implementation of these goals. Firstly, an audit of the industry's needs is being carried out to identify potential areas for improvement in terms of sustainability, involving various stakeholders in the process. Secondly, research results are being analysed to develop and implement eco-innovative technologies and methodologies in these areas of improvement, establishing a strong link between tourism and research, development and innovation (R&D and Innovation).

One of the crucial goals of this project is to promote the dissemination of the proposed eco-innovations, thus facilitating their deployment in the region's tourism industry. The ultimate goal is to formulate an effective strategy for the implementation of the circular economy.

With this objective in mind, a book has been compiled to address the various perspectives of the circular economy in the tourism industry. The publication consists of three main sections: 'Challenges and Opportunities', 'Good Practices', and 'Research, Innovation, Competitiveness and Production'. It explores the many aspects of the circular economy in tourism, showcases examples of good practice and discusses the implications of adopting the circular economy in terms of innovation, competitiveness and production.

The central purpose of this work is to create an effective link between theory and practice, making it easier for tourism businesses and stakeholders to effectively incorporate circular economy principles. It aims to provide a clear path towards a more sustainable, profitable and responsible future for tourism in the Valencian Region through practical recommendations, evidence-based research and real-life examples.

Despite the significant contributions made by tourism to economic development and job creation, it is often overshadowed by unsustainable practices. This book aims to address these concerns by proposing a transition from traditional linear production and consumption patterns to the circular economy. The focus is not only on reducing waste and optimizing resource use, but also on cultivating an organizational culture that promotes sustainable practices.

The first segment of the book, entitled 'Challenges and Opportunities', explores the journey of several tourism businesses in the Valencian Region towards the circular economy. It explores the five steps required to achieve circularity, from awareness to action, reflecting on how this can mitigate environmental and social impacts. It examines the relevance of an 'impact culture' and the role it plays in the transition to a circular economy, providing strategic recommendations for its implementation in the Valencian tourism industry.

The second section, 'Best Practices', delivers practical examples of the implementation of circular practices in various sub-sections of the tourism industry, including hotels and restaurants, as well as niche markets such as wine and beer tourism. Our researchers offer insights into the various ways in which companies can build circularity into their operating models, and we present case studies highlighting successful practices.

The third and final part, 'Research, Innovation, Competitiveness and Production', explores the potential benefits of adopting a circular economy model on a larger scale. This section explores the relationship between green innovation and corporate performance in the hospitality industry and suggests how a supportive organizational culture can drive the success of these innovations. We then discuss the application of circularity best practices in Spain, with an emphasis on the role of the hotel industry in promoting sustainable tourism. The section showcases strategies for balancing economic development with environmental conservation, while maintaining the quality of the tourism experience.

Throughout this book, our primary objective is to draw a connection between conceptual principles and their actual application, enabling businesses and stakeholders in the industry to effectively incorporate the principles of the circular economy. Through practical recommendations, evidence-based research and real-world examples, we hope to provide a roadmap to a more sustainable, profitable and responsible future in tourism.

Acknowledgements The project has received funding from the Valencian Innovation Agency (AVI) Complementary actions to promote and strengthen innovation 2021. Research Project INNACC/2021/49, "Creation of an Innovation Platform for the Promotion and Implementation of a Circular Economy Strategy in the Tourism Sector of the Valencian Region (InnoEcoTur)".

Open Access This chapter is licensed under the terms of the Creative Commons Attribution 4.0 International License (http://creativecommons.org/licenses/by/4.0/), which permits use, sharing, adaptation, distribution and reproduction in any medium or format, as long as you give appropriate credit to the original author(s) and the source, provide a link to the Creative Commons license and indicate if changes were made.

The images or other third party material in this chapter are included in the chapter's Creative Commons license, unless indicated otherwise in a credit line to the material. If material is not included in the chapter's Creative Commons license and your intended use is not permitted by statutory regulation or exceeds the permitted use, you will need to obtain permission directly from the copyright holder.

Part I
Challenges and Opportunities

Chapter 2
The Transition Journey to the Circular Economy by Companies in the Valencian Region's Tourism Industry

Blanca de-Miguel-Molina, María de-Miguel-Molina, Luis Miret-Pastor, and María Belén Silva-Cárdenas

Introduction

The transition to the circular economy involves three main principles: "eliminate waste and pollution, circulate products and materials, and regenerate nature" (Ellen MacArthur Foundation, 2022). The distinctiveness of the circular economy is that it enables the simultaneous improvement of at least seven Sustainable Development Goals (United Nations, 2015): SDG6 (Clean water and sanitation), SDG7 (Affordable and clean energy), SDG11 (Sustainable cities and communities), SDG12 (Responsible consumption and production), SDG13 (Climate action), SDG14 (Life below water), SDG15 (Life and land). This might provide reasons to understand why the European Commission established that the circular economy would be a prerequisite to the achievement of climate neutrality by 2050 in its Circular Economy Action Plan (2020). Likewise, the European Commission connected the circular economy to improvements in the Sustainable Development Goals.

When companies in the tourism sector decide to be more "circular", they reach achievements on this journey but also encounter difficulties. They have shared these with us during the Innoecotur project (https://innoecotur.webs.upv.es/).

In this chapter, we present our findings after conducting three focus groups with 18 executives from hotels and restaurants. The information has been divided into main five steps perceived after analysing managers' comments.

B. de-Miguel-Molina (✉) · M. de-Miguel-Molina · M. B. Silva-Cárdenas
Department of Business Organisation, Universitat Politècnica de València, Valencia, Spain
e-mail: bdemigu@omp.upv.es; mademi@omp.upv.es; mbsilcar@alumni.upv.es

L. Miret-Pastor
Department of Economics and Social Sciences, Universitat Politècnica de València, Gandia, Spain
e-mail: luimipas@esp.upv.es

© The Author(s) 2024
M. Segarra-Oña et al. (eds.), *Managing the Transition to a Circular Economy*, SpringerBriefs in Business, https://doi.org/10.1007/978-3-031-49689-9_2

The chapter is set out as follows. After this introduction, a literature review is conducted for the two main frameworks that we use to support the analysis of the information collected in the focus groups. Then, the five steps on the journey to circularity are explained with examples of comments made by participants. Finally, the main conclusions are revealed.

Frameworks in the Circular Economy Literature

Literature about the circular economy offers two supporting frameworks for companies on their journey to becoming more circular. These frameworks are the barriers to transition and the R concepts for circularity. Table 2.1 shows a summary of seven barriers obtained through a literature review, although only two of these works refer to the hospitality sector. It is important to note that barrier analysis has emerged as an important topic in the last two years.

The number of R concepts has risen during the last decade. In our literature review, we found papers focused on three Rs (Ioannidis et al. 2021; Westgeest, 2022), four Rs (Kirchherr et al., 2017), nine Rs (Reike et al., 2018, 2023) and up to sixty Rs (Uvarova et al., 2023). In spite of the variety of synonyms expressing R concepts, Kirchherr et al. (2017) stated that the four most common R concepts are reduce, reuse, recycle and recover. Table 2.2 shows the nine Rs defined by Reike et al. (2018). We have included some examples from the Marriott hotel group's ESG[1] report.

Focus Groups: Design

To analyse the journey of companies on their transition to the circular economy, information was collected through three focus groups conducted in Alicante, Castellon and Valencia. Every focus group involved six participants, who were managers with experience in the transition to the circular economy. Moreover, two of the participants were executives in hotels, two managed restaurants, and the other two led firms which supplied products to hotels and restaurants. Hotels were selected according to their category and had four stars or higher. In the case of restaurants, the criterion was being included in a ranking or having received an award.

A guide was drafted to steer the three focus groups to facilitate the comparison of data collected. The University of Alicante, Jaume I University and the University of Valencia oversaw the selection of participants, moderated the three focus groups and transcribed the three recordings.

---

[1] ESG: Environmental, Social, and Governance.

Table 2.1 Barriers in the transition to the circular economy

Barriers	Examples	Authors
Economic/Financial	Cost and financial barriers; Economically dominated thinking; Unwillingness to engage in trade-offs	Khan et al. (2022), Hina et al. (2022), Mishra et al. (2022), Takacs et al. (2022); Shao et al. (2023)
Competitive barriers	Low level of profit and market demand level; Consumer-related barriers; Market barriers; Perception that the quality of the finished product could be compromised; Product design	Khan et al. (2022), Hina et al. (2022), Mishra et al. (2022), Takacs et al. (2022), Shao et al. (2023)
Implementation barriers	The learning process and associated risk; Feasibility of circular economy implementation; Unused/wastage of material; Company's policies and strategies; Lack of industrial support; Collaboration; Complex and less integrated supply chain; Supply chain-related barriers; Staff motivation	Khan et al. (2022), Hina et al. (2022), Westgeest (2022), Shao et al. (2023)
Knowledge barriers	Lack of knowledge and skills	Khan et al. (2022), Mishra et al. (2022), Takacs et al. (2022)
Other resource barriers	Technical and/or technological barriers; Lack of physical and intellectual resources; Human resource shortages	Khan et al. (2022), Hina et al. (2022), Mishra et al. (2022), Takacs et al. (2022)
Cultural barriers	Top management resistance to change; Risk aversion; Strategic barriers; Internal stakeholders; Social, cultural and environmental barriers; Cultural barriers; Organisational barriers	Khan et al. (2022), Hina et al. (2022), Mishra et al. (2022), Takacs et al. (2022), Shao et al. (2023)
Administrative & Institutional barriers	Lack of government policies concerning circular economy; Legislative and economic barriers; Government and regulatory barriers	Khan et al. (2022), Hina et al. (2022), Mishra et al. (2022), Takacs et al. (2022), Westgeest (2022), Shao et al. (2023)

Source: Various

The analysis of the transcribed information was conducted by the Universitat Politècnica de València using the QDAMiner 5 software. Content analysis was the method selected to extract the information, code the sentences and divide them into the following groups:

1. Awareness about the transition to the circular economy.
2. Barriers and incentives in the transition to the circular economy.
3. Decisions for the transition to the circular economy.
4. Implementation of circular economy initiatives.
5. Reporting on advances in the transition to the circular economy.

The process followed in the content analysis resulted in 166 codes after a first draft and code reorganisation into groups.

Table 2.2 R concepts with examples

Rs	Examples (producer side)[a]	Example in Marriott[b]
R0. Refuse	Refuse the use of specific materials	Cage-free eggs
R1. Reduce	Using less material per unit of production	Switching single-use plastic in toiletry bottles to larger pump-topped bottles. Eliminate the use of paper for hotel reporting processes. Use of renewable energy. Responsible source
R2. Resell/Re-use	Enabling multiple re-uses	Laundry water reuse systems. Use food that would have been potentially discarded to create fresh jam from passion fruit skin
R3. Repair	Repair to extend the lifetime of the product	
R4. Refurbish	Components are replaced or repaired resulting in an upgrade of the product	MindClick's Design Impact Report™, to evaluate the environmental and social responsibility of new builds and renovations based on specified interior furnishings, operating supplies and equipment
R5. Remanufacture	Reconditioning, restoring	Reuse and breathe new life into existing land or buildings, rather than destroying old sites and rebuilding using new materials
R6. Repurpose	Reusing discarded goods or components adapted for another function	Send cleaned and repacked soap to local communities for hygiene needs
R7. Recycle materials	Separate and use secondary materials	Recycle used Nespresso capsules from guest rooms, returning one million capsules back to Nespresso
R8. Recover (energy)	Waste treatment to produce energy	In Hong Kong work with the city's first organic resources recovery centre, to convert food waste into electricity. Remove seaweed from beaches and sequester and store carbon in the soil
R9. Re-mine	Reprocessing	

Source: [a]Reike et al. (2018); [b]Marriott International (2022)

Focus Groups: Results

The results are presented according to the five steps that tourism companies follow on their journey to the circular economy. Figure 2.1 shows a summary of these steps and the main topics involved in them. The five steps are the five code groups indicated in the previous section. Therefore, the first step starts when a company considers the transition to the circular economy, while the second step refers to the barriers, the third step to the decisions they make about what initiatives to undertake, the fourth step focuses on the implementation of these initiatives, and the fifth step involves companies' reporting advances in terms of the circular economy.

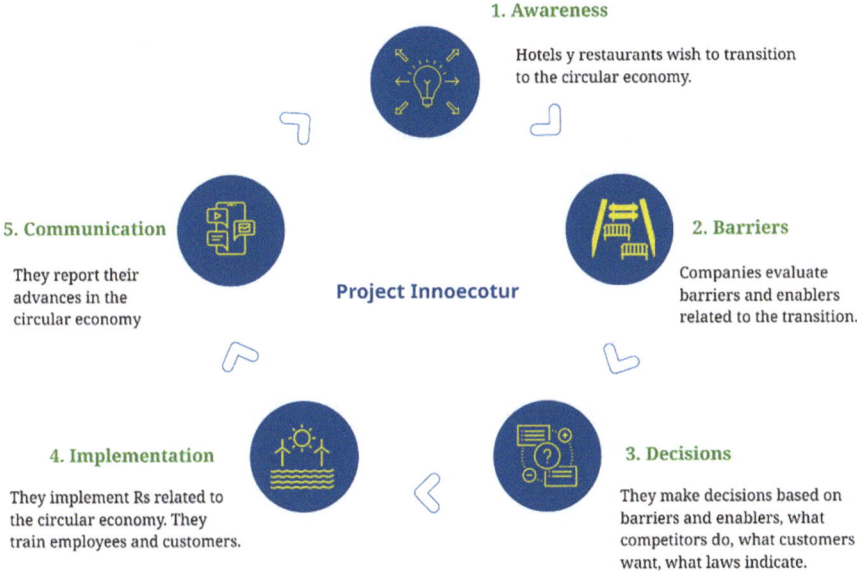

Fig. 2.1 The transition journey to the circular economy

Table 2.3 Frequency of codes about awareness

The circular economy is important	Hotels	Restaurants	Suppliers	Total	%
A1_For everything and everybody	1	1	2	4	22%
A2_ For customers	2	2	4	8	44%
A3_For companies	2	2	4	8	44%
A4_For employees and customers	1	1	2	4	22%
A5_For global agencies	1	1	2	4	22%

Awareness About the Transition to the Circular Economy

This group of codes includes the opinions of participants in the three focus groups about the reasons why their companies considered transitioning to the circular economy. Participants indicated the importance for their own companies, customers and international organisations, some of which had to include sustainability activities to participate in contracts. Table 2.3 shows a summary of codes related to awareness. The frequencies indicate that participants believed that the interest of companies to transition to the circular economy was that customers demanded sustainability, and this becomes the purpose of companies.

Examples that illustrate the comments behind the codes in Table 2.3 include the following A2.1 and A2.2 are examples for code A2 in Table 2.3, while A3, A4 and A5 are linked to their corresponding codes in the table.

- A2.1. The circular economy is important to customers.
- A2.2. We have won a big client and the reason why was that we obtained a sustainability certification. We are working to reduce our carbon footprint by incorporating photovoltaic energy, for example.
- A3. More and more companies are working on the transition to the circular economy, inside and outside companies.
- A4. Circular economy initiatives will be an important factor to attract talent and customers to a hotel and other businesses.
- A5. Many international agencies include sustainability initiatives as a prerequisite for companies interested in participating in tenders.

Barriers and Enablers to Implementing the Circular Economy in the Tourism Industry

On the second step in the transition to the circular economy, companies in the tourism industry evaluate the potential barriers and enablers they might find when they implement actions to become more circular (Tables 2.4 and 2.5).

The opinions of participants about barriers to implementing the circular economy were divided into six types: economic, business, knowledge, cultural, social,

Table 2.4 Frequency of codes about barriers in the transition to the circular economy

Barriers	Hotels	Restaurants	Suppliers	Total	%
B1_Economic barriers	1	1	2	4	22%
B2_Business barriers	1	1	2	4	22%
B3_Knowledge barriers	2	2	4	8	44%
B4_Cultural Barriers	3	3	6	12	67%
B5_Social barriers	1	1	2	4	22%
B6_Administrative barriers	2	2	4	8	44%
B7_Institutional barriers	1	1	2	4	22%

Table 2.5 Frequency of codes for incentives to companies

Incentives available	Hotels	Restaurants	Suppliers	Total	%
I1_Lack of incentives	2	2	4	8	44%
I2_There are incentives	1	1	2	4	22%
I3_Companies should not be penalised when they do the right thing	2	2	4	8	44%
I4_Reward companies which do the right thing	2	2	4	8	44%
I5_Lack of information	2	2	4	8	44%
I6_Incentives should exist throughout the value chain	1	1	2	4	22%
I7_Incentives are welcome	1	1	2	4	22%

administrative and institutional. Table 2.4 shows the number of participants who indicated the presence of each barrier by sector. Frequencies indicate that cultural barriers are the most important in the tourist industry.

Economic barriers refer to the cost associated with measures that help companies in their transition to the circular economy. Participants in the focus groups indicated that the investments needed for the transition imply elevated costs which are difficult to take on board. Examples of participants' comments about code B1 in Table 2.4 were:

- B1.1. The trade-off between sustainability and profit is when it is difficult to ensure profitability and a return on investment.
- B1.2. The company cannot afford the cost involved in implementing an improvement, like switching to electric vehicles.
- B1.3 There is a cost related to recovering the containers and sending them to other companies.

Business barriers express the risk of eroding competitiveness when the cost associated with implementing actions to be "circular" results in lower margins than those of competitors. Examples of participants' comments about code B2 for this barrier in Table 2.4 were:

- B2.1. Being "circular" implies investing a lot of time to find suppliers as there is no list of circular suppliers available. The time invested in the process of finding them increases the cost for the company.
- B2.2. Reducing our margins due to investments in circularity makes it difficult to compete with rivals which do not invest and do not even know what laws demand.

Knowledge barriers are about the lack of knowledge related to what the circular economy is. Participants' opinions about code B3 indicated that this barrier appears with customers, companies and governments:

- B3.1. There is a need to translate the circular economy into a language that everybody understands, with examples indicating how many euros a company might save, as occurred with LED lightbulbs.
- B3.2. Customers do not know what sustainable gastronomy, ecological, and proximity products are.

Cultural barriers deal with the resistance to change, in this case, to transition to the circular economy. This barrier is present in both companies and customers. Participants in the focus groups mentioned the following about code B4:

- B4.1. Sometimes you have to fight against those who do not understand or who only see it as an extra cost.
- B4.2. When the customer does not demand sustainable products.
- B4.3. Risk aversion and being unable to work in a different way than we are used to.

Social barriers concern the engagement of all the people in the transition to the circular economy, especially those who are more aware of sustainability. This code in B5 was expressed by a woman who defended the involvement of other women because they are more interested in the circular economy:

- B5. Give opportunities to women as they are more aware of the circular economy.

Administrative and institutional barriers deal with laws and regulations which limit companies' progress towards the circular economy. Examples of opinions related to codes B6 and B7 were:

- B6. Local government barriers block the implementation of innovative waste-reducing initiatives.
- B7. Additional benefits could be given to companies to encourage the efforts they have made towards the circular economy.

Enablers in the form of incentives to companies include codes that refer to participants' comments about their perceptions of the availability or lack of these incentives. They also felt that companies are sometimes penalised despite doing the right thing. Table 2.5 includes the codes I1 to I7, defined for the analysis of incentives.

Examples of comments cited by participants in relation to code I5's lack of information show their different perceptions depending on the sector:

- I5.1. There is a lack of information when the sector is fragmented with many small companies, i.e., the restaurant sector.
- I5.2. We obtain all the information available because the Hotels Association in our region is a strong organisation.

Decisions About the Transition to the Circular Economy

When companies evaluate their options to transition to the circular economy, they consider what is valuable about what they offer to customers. This is because companies adapt their products and services to what customers really want. However, even if customers are aware of the circular economy, companies need to confirm that their suppliers are also willing to transition to the circular economy. Table 2.6 includes the frequency of codes about decisions related to the circular economy. The data in the table indicates that the transition might be a challenge for some companies based on customers' interest in the circular economy.

Examples of comments around the codes mentioned by participants in the focus groups, which are shown in Table 2.6 are:

- D5.1. Sometimes cultural barriers appear on the suppliers' side, as they do not accept the idea of moving towards the circular economy or obtaining a certification.

Table 2.6 Frequency of codes for decisions for the transition to the circular economy

	Hotels	Restaurants	Suppliers	Total	%
D1. Customers are not aware of the circular economy	3	3	6	12	67%
D2. Customers are aware of the circular economy	3	3	6	12	67%
D3. Some customers are more much aware of the circular economy than others	2	2	4	8	44%
D4. We do select suppliers based on their advances towards the circular economy	2	2	4	8	44%
D5. Some suppliers do not accept the idea of moving towards the circular economy	2	2	4	8	44%
D6. Suppliers accept our demands to moving towards the circular economy	1	1	2	4	22%
D7. Lack of a well-organised supplier network	1	1	2	4	22%
D8. It is difficult to find suppliers for some products	1	1	2	4	22%

Table 2.7 Frequency of codes for the implementation of actions

Implementation of circular economy initiatives	Hotels	Restaurants	Suppliers	Total	%
Measure before reducing	3	3	6	12	67%
Sustainable Development Goals as a guide	1	1	2	4	22%
R0. Refuse	1	1	2	4	22%
R1. Reduce	3	3	6	12	67%
R2. Reuse	2	2	4	8	44%
R5. Remanufacture	1	1	2	4	22%
R7. Recycle	1	1	2	4	22%
R8. Recover	3	3	6	12	67%

– D6.1. Suppliers accept our demands if the investment required to move towards the circular economy is not substantial.
– D7.1. The lack of a well-organised supplier network implies that we need to spend a lot of time selecting products day after day.

Implementation of Circular Economy Initiatives

This step involves the implementation of initiatives that companies have selected after considering barriers and enablers, what customers really want, suppliers' capabilities, and the impact on profitability. Table 2.7 shows the frequencies for R concepts applied by companies participating in the focus groups. To organise the information collected, the nine Rs model by Reike et al. (2018) was applied.

Kirchherr et al. (2017) concluded that the four most common R concepts were: reduce, reuse, recycle and recover. In our results, the most frequently mentioned concepts were reduce, recover and reuse.

Examples of comments on R codes included in Table 2.7 were:

- R0.1. Plastic bags from laundry suppliers are refused, and reused textile bags are the norm. Boxes from suppliers are refused.
- R1.1. The surplus energy from photovoltaic infrastructures is used in the swimming pool in some months of the year.
- R2.1. Water from showers and water tanks is recovered and reused to water the golf course. Additionally, water from showers is used in the water tanks.
- R5.1. The building was refurbished instead of constructing a new one.
- R7.1. We recycle plastic and create products with it.
- R8.1. Plastic is recovered to prevent it being thrown into the dustbin.

It is clear from the participants' comments that waste management is an important element in the circular economy. In fact, we obtained additional information from participants about how they reduce waste. Some comments were coded as follows:

- W1. Managing waste (22% companies): W1.1. We separate waste in specific containers. W1.2. We work with waste management companies that support social causes.
- W2. Challenges with waste management (22% companies): W2.1. There is a need to evaluate where waste is generated in the value chain. We end up paying costs which clearly belong to our suppliers. W2.2. Separating waste and recycling costs money.

The transition to the circular economy also requires training employees and customers to increase their awareness of circularity. Two codes were defined to indicate the two recipients of training, employees and customers:

- T1. We train employees about the circular economy (22% companies).
- T2. We train customers about the circular economy (22% companies).

Reporting Circular Economy Initiatives

Reporting advances in circularity is the last step on the journey towards the circular economy. The defined codes indicate what companies report, and they indicate that companies consider the circular economy as adding value for customers and as part of their experience:

- C1.1. We report the value of the circular economy initiatives we implement (22.2% companies).
- C1.2. We report the customer experience about the circular economy initiatives we implement (22.2% companies).

- C2.1. We report the circular economy certification obtained (22.2% companies).
- C2.2. We report our carbon footprint in real-time (22.2% companies).

Conclusions

This chapter presents the results of a study undertaken to map the journey taken by tourism companies in their transition to the circular economy. The analysis centred on the tourism industry in the Valencian Region and was based on the information collected from three focus groups, with industry managers who have experienced a transition to the circular economy. The analysis was divided into the five steps on companies' transition journey. The main conclusions obtained from the analysis of the different steps on the journey were the following.

The conclusion from the awareness step is that companies basically transitioned to the circular economy because they were customer-centred and, therefore, they tried to offer circularity because this is what customers demanded.

The barriers and enablers step revealed that the most important barrier was cultural according to the companies. This manifested itself as an aversion to risk as the transition to the circular economy implies doing things in a different way. This barrier also applied to suppliers, denoting a challenge for companies when there is no well-organised supplier network that can offer sustainable products. In terms of enablers, companies mentioned that they sometimes felt as though they were penalised despite doing things well.

The main conclusion of the decisions step referred to the impact of cultural barriers in terms of suppliers. The result is that, although customers are aware of the circular economy, companies cannot provide a circular experience to them because of constraints imposed by suppliers.

The main conclusion of the implementation step was that the advances in the R concepts made by companies in the sector analysed centred on actions that involve reducing, recovering and reusing. The analysis of information provided by participants in the focus groups indicated that companies implemented the same initiatives as the world's leading hotel groups.

On the last step, reporting the initiatives implemented by companies, the main conclusion is that companies reported both the added value for customers and the certifications that prove this, for example carbon footprint levels in real-time.

References

Ellen MacArthur Foundation. (2022). *Circulytics. Definitions*. San Francisco: MacArthur Foundation. Retrieved from https://ellenmacarthurfoundation.org/resources/circulytics/resources

European Commission. (2020). *A new circular economy action plan for a cleaner and more competitive Europe*. European Commission, Brussels. Available in https://eur-lex.europa.eu/legal-content/EN/ALL/?uri=COM:2020:98:FIN

Hina, M., Chauhan, C., Kaur, P., Kraus, S., & Dhir, A. (2022). Drivers and barriers of circular economy business models: Where we are now, and where we are heading. *Journal of Cleaner Production, 333*, 130049.

Ioannidis, A., Chalvatzis, K. J., Leonidou, L. C., & Feng, Z. (2021). Applying the reduce, reuse, and recycle principle in the hospitality sector: Its antecedents and performance implications. *Business Strategy and the Environment, 30*(7), 3394–3410.

Khan, S. A., Mubarik, M. S., & Paul, S. K. (2022). Analyzing cause and effect relationships among drivers and barriers to circular economy implementation in the context of an emerging economy. *Journal of Cleaner Production, 364*, 132618.

Kirchherr, J., Reike, D., & Hekkert, M. (2017). Conceptualizing the circular economy: An analysis of 114 definitions. *Resources, Conservation and Recycling, 127*, 221–232.

Marriott International. (2022). *Serve 360 Report: Environmental, social, and governance progress.* Available in http://serve360.marriott.com/wp-content/uploads/2022/10/Marriott-2022-Serve-360-ESG-Report-accessible_F.pdf

Mishra, R., Singh, R. K., & Govindan, K. (2022). Barriers to the adoption of circular economy practices in micro, small and medium enterprises: Instrument development, measurement and validation. *Journal of Cleaner Production, 351*, 131389.

Reike, D., Hekkert, M. P., & Negro, S. O. (2023). Understanding circular economy transitions: The case of circular textiles. *Business Strategy and the Environment, 32*(3), 1032–1058.

Reike, D., Vermeulen, W. J., & Witjes, S. (2018). The circular economy: new or refurbished as CE 3.0?—Exploring controversies in the conceptualization of the circular economy through a focus on history and resource value retention options. *Resources, Conservation and Recycling, 135*, 246–264.

Shao, J., Aneye, C., Kharitonova, A., & Fang, W. (2023). Essential innovation capability of producer-service enterprises towards circular business model: Motivators and barriers. *Business Strategy and the Environment, 32*, 4548–4567.

Takacs, F., Brunner, D., & Frankenberger, K. (2022). Barriers to a circular economy in small-and medium-sized enterprises and their integration in a sustainable strategic management framework. *Journal of Cleaner Production, 362*, 132227.

United Nations. (2015). *Transforming our world: the 2030 Agenda for sustainable development.* General Assembly on 25 September 2015, A/RES/70/1. Available in https://sdgs.un.org/2030agenda

Uvarova, I., Atstaja, D., Volkova, T., Grasis, J., & Ozolina-Ozola, I. (2023). The typology of 60R circular economy principles and strategic orientation of their application in business. *Journal of Cleaner Production, 409*, 137189.

Westgeest, L. (2022). A circular economy approach to plastics in hospitality: A Frisian case study. *Research in Hospitality Management, 12*(3), 299–308.

Open Access This chapter is licensed under the terms of the Creative Commons Attribution 4.0 International License (http://creativecommons.org/licenses/by/4.0/), which permits use, sharing, adaptation, distribution and reproduction in any medium or format, as long as you give appropriate credit to the original author(s) and the source, provide a link to the Creative Commons license and indicate if changes were made.

The images or other third party material in this chapter are included in the chapter's Creative Commons license, unless indicated otherwise in a credit line to the material. If material is not included in the chapter's Creative Commons license and your intended use is not permitted by statutory regulation or exceeds the permitted use, you will need to obtain permission directly from the copyright holder.

Chapter 3
Impact Culture and the Circular Economy in the Tourism Industry: An Analysis of Challenges and Recommendations for Sustainability

Virginia Santamarina-Campos, Miguel Ángel Mas-Gil, María de-Miguel-Molina, and Daniel Catalá-Pérez

Introduction

Tourism is an essential part of the economy in many regions around the world, as it is a major driver of growth. However, as environmental and social concerns gain traction, there is a pressing need to address its impacts. The circular economy, which prioritises resource efficiency, waste reduction and reuse, is an opportunity to transform the sector and move towards more sustainable, responsible models.

The circular economy is defined as an economic model in which resources are used efficiently, minimising waste generation and maximising recycling and reuse. This approach implies a shift in the way we produce and consume towards a more sustainable, responsible model. Implementing the circular economy in the tourism industry can generate substantial environmental and economic benefits. According to the European Commission's circular economy in tourism report, these benefits include reducing greenhouse gas emissions, creating green jobs, improving competitiveness and reducing costs (Einarsson & Sorin, 2020). In addition, according to the World Travel & Tourism Council's (WTTC) study, the circular economy can help tourism businesses to reduce costs and improve their long-term sustainability, while mitigating their environmental footprint (World Travel & Tourism Council & Harvard T.H. Chan, 2022).

V. Santamarina-Campos (✉)
Department of Conservation and Restoration of Cultural Assets, Universitat Politècnica de València, Valencia, Spain
e-mail: virsanca@upv.es

M. Á. Mas-Gil · M. de-Miguel-Molina · D. Catalá-Pérez
Department of Business Organisation, Universitat Politècnica de València, Valencia, Spain
e-mail: mimagi@omp.upv.es; mademi@omp.upv.es; dacapre@ade.upv.es

© The Author(s) 2024
M. Segarra-Oña et al. (eds.), *Managing the Transition to a Circular Economy*, SpringerBriefs in Business, https://doi.org/10.1007/978-3-031-49689-9_3

However, its implementation also poses challenges, such as lack of knowledge and training, dependence on external factors (for example, local infrastructure and business cooperation), and the need for cultural and behavioural changes among consumers and tourism businesses. In this sense, impact culture plays a crucial role in driving the adoption of more sustainable and responsible behaviours. According to Clark (2014), impact culture refers to the adoption of behaviours and practices that generate positive, sustainable change, and its scarcity could contribute to the lack of innovation and redesign which has been identified in the Valencian Region.

In 2021, Spain took significant steps towards sustainability in tourism. The Ministry for the Ecological Transition and the Demographic Challenge launched a rural recovery plan with a series of initiatives and measures aimed at combating depopulation and boosting the development of sustainable destinations in rural areas, thus promoting the circular economy (Ministerio para la Transición Ecológica y el Reto Demográfico, 2021). Likewise, the Plan to modernise and boost competitiveness in the tourism sector intends to mobilise billions of euros to make the industry more competitive, promoting sustainability and the circular economy (Gobierno de España, 2021). Moreover, new trends that have emerged after the Covid-19 pandemic have generated a growing interest in experiences and destinations linked to health and wellness, with the circular economy being one of the main trends in wellness tourism (Instituto Valenciano de Tecnologías Turísticas, 2022).

This chapter presents the results of a Participatory Action Research (PAR) process carried out in the provinces of Castellon, Valencia and Alicante in 2022, as part of the InnoEcoTur project. The aim of this study was to define strategic recommendations for the implementation of the circular economy in the tourism industry. In this context, we underline the relevance of culture in driving the transition towards the circular economy.

The Valencian Region's Strategic Tourism Plan highlights the importance of the circular economy for the sustainability and competitiveness of the sector and promotes its implementation through efficient waste management, resource use and the promotion of local, seasonal products and services (Secretaría Autonómica de Turismo & Instituto Valenciano de Tecnologías Turísticas, 2020). It highlights the need to forge robust alliances between the different tourism stakeholders to enhance the circular economy and promote sustainability. Given the huge potential for the implementation of circular practices in the industry, the emphasis is placed on the need for a culture that encourages and supports this transition. The Strategic Plan and the Tourism Sustainability Plan rolled out by Visit València (2022) run in this direction and are committed to promoting the circular economy. These plans include measures for its implementation, such as recycling and reuse of waste, the promotion of sustainable products and services and the efficient management of water and energy resources.

Participatory Action Research Methodology with Tourism SMEs to Facilitate Their Transition to the Circular Economy

This study applies the PAR methodology to analyse the needs and challenges of tourism SMEs in their transition towards the circular economy. This method, which is widely used in the tourism sector and in circular economy projects, involves stakeholders in the research process. A systemic approach and multi-stakeholder collaboration are crucial to kickstart the green, circular recovery process focusing on environmental, social and economic dimensions (Einarsson & Sorin, 2020).

Three participatory sessions were held in Castellon, Valencia and Alicante in 2022. We used a scenario matrix and categorised flashcards to engage participants. The cards were divided into groups based on the Sustainable Development Goals (SDGs), actions for the circular economy (European Commission, 2020), the seven Rs of the circular economy (Arisi, 2020; Capgemini Research Institute, 2021), and previously identified barriers and measures (de Miguel Molina et al., 2022) (see Figs. 3.1, 3.2). The scenario matrix is a tool that encourages reflection and elicits information from participants about the present and the foreseeable/desired future of the circular economy. The toolkit included two large magnetic boards and a set of

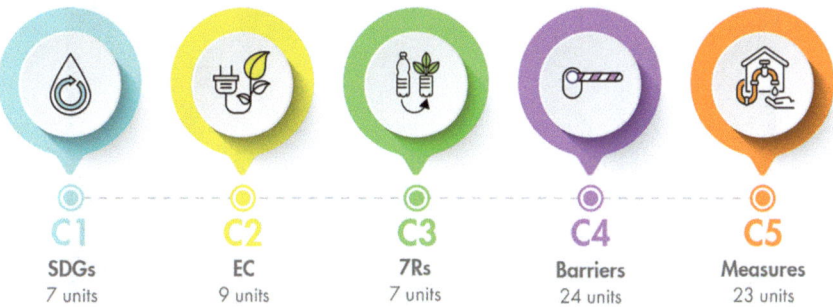

Fig. 3.1 Flashcard categories. Source: authors' own, 2023

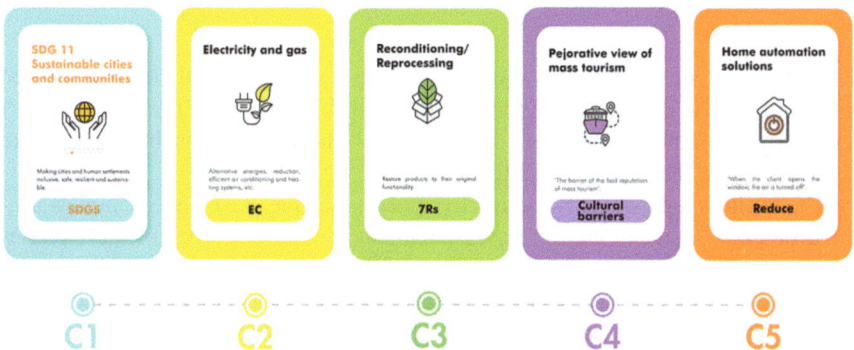

Fig. 3.2 Example of flashcards in the five categories. Source: authors' own, 2023

Fig. 3.3 Stages of implementation of the methodology. Source: Own work, 2023

Fig. 3.4 Detail of the presentation phase of the PAR methodology held in Castellon, in June 2022. Source: InnoEcoTur Project, 2022

70 flashcards. One of the boards showed the cards by category, and the other showed a scenario matrix design, divided into four sections: "This doesn't exist and I don't want it to exist", "This exists but I don't like it", "This doesn't exist but I would like it to exist" and "This exists and I like it".

The methodology was divided into five phases (see Fig. 3.3): presentation, warm-up, group development, group reassessment and conclusions. Each of these phases was carefully planned to ensure the quality of the results.

1. Presentation: The moderator introduced the topic, contextualising the exercise in the framework of the project, and explained the objective and purpose of the activity (see Fig. 3.4).

Fig. 3.5 Detail of the warm-up phase of the PAR methodology carried out in Valencia in July 2022. Source: InnoEcoTur Project, 2022

2. Warm-up: Two large magnetic boards were used, one with the scenario matrix and the other with the flashcards. The moderator explained to the participants how to complete the matrix collaboratively using the cards, but without interacting with each other (see Fig. 3.5).

3. Group development: Participants looked at the final matrix output and discussed possible changes, explaining the rationale for their placement of each flashcard (see Fig. 3.6).

4. Group reassessment: Once the group had reached a consensus on the scenario matrix, participants worked together to design strategies to promote the application of circular economy principles. This joint analysis was performed using the matrix and with an online template based on the matrix (see Fig. 3.7). The template addressed five specific questions (see Fig. 3.8), including aspects such as the need to change behaviours, the implementation of a new production model and the adoption of eco-design.

5. Conclusions: The moderator gave a short summary of the activity, thanking everyone for their participation.

This methodological approach may limit comparisons between the three sessions as there was a lack of absolute control over the activity. However, this limitation was reduced by applying a uniform design and material. It also had the constant support of the same staff in all the sessions, thus ensuring the consistency of the process in each location.

Furthermore, participant selection was based on both homogeneity and intra-group heterogeneity criteria. The participants, who represented diverse discursive perspectives in relation to the case study, were divided into five different groups. These groups were made up of staff from small and medium-sized enterprises

Fig. 3.6 Detail of the group development phase of the PAR methodology in Valencia, July 2022. Source: InnoEcoTur Project, 2022

(SMEs) in the following categories: hospitality, catering, hotel suppliers, and restaurant suppliers, as well as civil servants working in the tourism industry.

We used our own video recordings and photographs of the activities for the data analysis.

Subsequently, the information from the templates and from the data extracted from the recordings on the distribution and movements of the flashcards in the matrix was compiled, considering the sequence of selection, individual and group distribution of the cards in the matrix, and reiterations from the individual to the group phases. The term 'reiteration' refers to cases where flashcards move into a different quadrant of the matrix between the individual and group phases, indicating a repetition of card selection or distribution. Therefore, flashcards with 'no reiterations' are those that have not moved from the individual to the group phases.

Qualitative and quantitative data analysis was carried out on the information collected. Various analytical techniques were used, including statistical functions, pivot tables and graphs. These tools enabled data to be processed for the exploration of patterns, as well as an analysis of variance.

This chapter focuses on the results from the group work phases (phases 3 and 4), as the data analysis showed that (across all three groups) 90% of the flashcards remained in the same quadrant of the matrix in the individual and group phases.

Fig. 3.7 Detail of the reassessment phase of the PAR methodology group in Alicante, June 2022. Source: InnoEcoTur Project, 2022

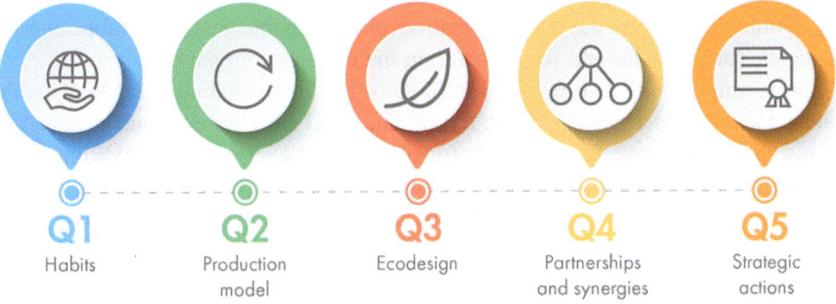

Fig. 3.8 Specific questions raised in relation to the strategic recommendations. Source: authors' own, 2023

Results

During the collaborative review of the scenario matrix and the group activity, we identified both strengths and weaknesses in the implementation of the circular economy and the SDGs in the Valencian Region's tourism industry. This analysis was necessary to develop a circular economy culture, as it provided a detailed diagnosis of the current situation to identify areas for improvement.

A favourable perception was observed for the implementation of and support for SDG 15 'Life on land' and SDG 13 'Take urgent action to combat climate change and its impacts', both of which are fundamental to create a culture that seeks to generate positive and significant changes in society and the environment. These SDGs are closely linked to the strategy of 'Reduce', 'Return/Recover' and 'Reuse', which are basic principles to combat pollution caused by plastics, textiles and household products.

In relation to resource extraction and use, SDGs 13 and 15 seek to mitigate negative environmental impacts, such as deforestation, desertification and land degradation, and to promote sustainable agricultural and fishing practices. Additionally, they aim to encourage sustainable travel, and increase the circular potential of batteries to reduce the environmental impact of transportation and improve hazardous waste management. In summary, the main objective of SDGs 13 and 15 is to promote sustainable, circular practices in resource use and environmental management to achieve global sustainable development. Given the Valencian Region's geographical location on the Mediterranean coast and its warm climate, there may be a positive perception and greater sensitivity towards the implementation of SDGs 13 and 15 in the region.

However, a lack of focus on the implementation of SDG 7 'Ensure access to affordable, safe, sustainable and modern energy for all' and SDG 11 'Make cities and human settlements inclusive, safe, resilient and sustainable' was detected, related to the absence of measures to obtain energy from waste, install sustainable lighting, reduce the carbon footprint and improve energy efficiency. Although the Valencian Region has a warm, sunny climate for much of the year, it seems that insufficient attention is given to energy efficiency and sustainable lighting and there are significant gaps in the implementation of sustainable and renewable energy strategies, such as solar and wind power, in the region's tourism industry.

Shortcomings were also detected in SDG 6 'Ensure availability and sustainable management of water and sanitation for all' and SDG 14 'Conserve and sustainably use oceans, seas and marine resources for sustainable development'. Additionally, a lack of focus on water-saving measures was identified, which is key to a coastal region such as the Valencian Region. Although the region has access to the sea and abundant water resources, it faces challenges in terms of sustainable water management, which has a significant relationship with SDG 6. The region is also a major agricultural hub, which implies high demand for water. It is also a tourist area, which implies higher water consumption in hotels, restaurants and other tourist establishments. In addition, the Valencian Region is closely linked to SDG 14 as it is a coastal region that relies heavily on tourism and fishing, and both sectors have a major impact on the marine environment. The region faces major challenges such as overfishing, water pollution and the degradation of marine ecosystems which significantly impact both environmental sustainability and the local economy. Therefore, the implementation of measures to conserve and sustainably use marine resources is essential to achieve sustainable development in the Valencian Region.

Furthermore, the lack of focus on SDG 12 'Ensure sustainable consumption and production patterns' is directly related to shortcomings in 'Recycle', 'Refurbish/

Reprocess', 'Repair/Rehabilitate and Redesign' strategies, plastic footprint reduction, zero landfill and circularity in building life cycles. These strategies aim to promote resource efficiency and encourage the adoption of sustainable practices in both production and consumption. These failures are indicative of the need to cultivate a stronger impact culture in the Valencian Region. The identification of these needs may be related to the fact that tourism is a major driver in the region, representing 15.5% of GDP (Turisme Comunitat Valenciana, 2022). Given that the tourism industry generates significant amounts of waste and consumes many resources, it is essential to adopt sustainable and circular practices in waste management, plastic footprint reduction and efficient resource use.

SDG 12 can also be related to the lack of measures focused on reducing but also on enhancing local products and agroecology. This SDG aims to promote the efficient use of resources and encourages the adoption of sustainable practices in production and consumption. However, the lack of attention to this aspect may negatively affect the promotion of agroecology and local products, which are an important source of sustainable development in the region. In addition, these shortcomings may lead to growing dependence on imported products, which may not meet the necessary sustainability and quality standards.

There was also a notable lack of measures targeting IT tools, the implementation of home automation solutions, electronics and Information and Communication Technologies (ICT), as well as a poor innovative culture. This situation is also connected to the lack of focus on SDG 12, which seeks to promote the adoption of sustainable consumption and production practices and the responsible use of resources and energy efficient technologies and processes. In this context, the lack of a strong impact culture in the Valencian Region could influence this gap.

The identification of weaknesses in the implementation of SDGs 7, 11, 6, 14, and 12 highlights the areas where the impact culture needs to be strengthened. These findings suggest that, despite positive efforts in some areas, the Valencian Region must continue to work to ensure that its tourism sector moves forward in a truly sustainable and environmentally friendly manner.

The dichotomous perception about reducing the industry's plastic footprint, which is seen as a strength and weakness, could be related to the importance of the plastics and rubber industries in the Valencian Region. These sectors come second in the national ranking (Hervas-Oliver et al., 2018) and, in fact, 21% of active Spanish rubber manufacturers are located in the Valencian Region (Instituto Valenciano de la Competitividad Empresarial, 2021).

The dualistic view on energy efficiency, seen as a strength (SDG 13 and SDG 15) and a weakness (SDG 12), could be influenced by the geographical location of the Valencian Region, characterised by a warm, sunny Mediterranean climate most of the year. Although the region has high renewable energy potential in the areas of solar and wind power, the lack of focus on SDG 12 could have a negative effect on the attention paid to energy efficiency and sustainable lighting. This could also be related to the lack of an integrated, holistic view of sustainability in tourism planning and natural resource management, which addresses environmental, social and economic challenges in a balanced way. Tourism planning may be focused on maximising short-term economic benefits, which can lead to a limited, binary approach to sustainability. Furthermore, the fact that sustainable agricultural practices are seen as a strength, while the lack of measures focused on reducing whilst boosting local products and agroecology is perceived as a weakness may indicate a lack of understanding of the links between the different aspects of sustainability and how to address them holistically.

The lack of gender perspective in circularity is identified in the quadrant "This doesn't exist, but I would like it to exist", which indicates a positive attitude. However, this perception may be conditioned by the use of double negation in the statement, creating confusion, given that the data indicates the presence of vertical and horizontal inequalities in the Valencian Region's tourism industry (Alonso-Monasterio Fernández, 2019).

The flashcards that ranked highest in the overall order of the three activities related to SDGs 14, 11 and 12, which also address the lack of an innovative, circular culture, a lack of consumer awareness, recycling, and rigid legislation in managing food waste and plastics. Therefore, priority was given to flashcards that reflected negative or problematic perceptions, as they seek to raise awareness and promote solutions to address the identified challenges. In all three activities, the flashcards reflecting a lack of innovative culture appeared higher than ninth position, indicating concern about the lack of innovation in the Valencian Region's tourism industry.

All SDGs, except SDG 13, appeared in the top 20 flashcards, demonstrating the relevance of these goals in the tourism sector. The fact that SDG 13 appeared among the last 20 flashcards may be due to the positive perception of its implementation in the region. On the other hand, SDG 14 appeared before the tenth flashcard in all three activities, which reflects the special concern in the Valencian Region given to the protection of the marine environment. Given that it is a coastal region with a major tourist industry, the preservation and sustainable management of marine resources is fundamental to guarantee the availability of these resources in the long term and to maintain the quality of tourist attractions. In conclusion, the concern for SDG 14 in the Valencian Region is related to the need to preserve and sustainably manage marine resources in a context of coastal tourism activity.

The PAR process revealed some consistency in participants' opinions when collaboratively reviewing the distribution of the flashcards in the matrix, although the quadrants "This exists but I don't like it" and "This doesn't exist but I would like it to exist" generated more discussion and reflection. This could also be conditioned by the direct effects of the flashcard colours, as there is a correlation between the average flashcard output as a function of the psychological and physiological effects of the colour of each category (Elliot, 2015; Heller, 2004). Group discussion proved useful in fostering cohesion and collaboration around common goals, especially on critical issues that require further attention in future studies.

During the group reassessment, strategic recommendations for implementing circular economy principles in the tourism sector were compiled. The responses once again emphasised the need for a cultural shift towards increased awareness and responsibility about resource utilisation. The importance of efficient waste and raw material management, the use of sustainable materials and spaces, and the implementation of measures to reduce waste in products and services were also highlighted. Furthermore, the importance of establishing alliances and synergies between different stakeholders, using the Quintuple Helix approach to drive the transition towards a circular economy in the tourism sector, was emphasised. It is essential to prioritise training and awareness-raising among all actors involved and ensure effective communication to promote the circular economy in tourism.

In short, these findings indicate that, although there are positive signs of an impact culture in the Valencian Region, there are still important challenges that need to be addressed to ensure that the region's tourism industry develops in a truly sustainable and environmentally friendly way. This involves working to strengthen the implementation of the SDGs and circular economy strategies, as well as fostering innovation and digital transformation.

Therefore, the results and group dynamics used helped to identify gaps and opportunities in the implementation of the circular economy and the SDGs in the tourism industry, which is useful for future studies and initiatives in this area. It is important to pay more attention to deficiencies and to promote greater consumer understanding and awareness, as well as the implementation of measures and strategies for more sustainable, responsible production in the tourism industry.

Conclusions

This collaborative study identified strengths and weaknesses in the implementation of the circular economy and the SDGs in the Valencian Region. While good progress was noted in some SDGs and strategies, significant gaps were also detected, such as water management, access to sustainable energy, waste management and sustainable production.

The study highlighted a lack of understanding of circularity and a deficient innovation culture in the industry. However, during the reassessment, key strategies to enhance the principles of the circular economy in tourism were identified, emphasising the need for a cultural shift towards greater awareness and responsibility in the use of resources, waste management, the use of sustainable materials and spaces, and the reduction of waste generated by products and services.

Furthermore, the importance of forging alliances and synergies between different stakeholders to promote the transition towards a circular economy was emphasised. In this regard, impact culture plays a crucial role in fostering changes in attitudes and behaviours and in driving the adoption of more sustainable practices.

In conclusion, this study highlights gaps and opportunities in the implementation of the circular economy and the SDGs in the tourism industry, providing valuable insights for future studies and initiatives. The findings underline the need to improve deficient areas, educate and raise consumer awareness, implement strategies for more sustainable, responsible production, and promote an impact culture. It highlights the importance of a holistic vision that balances environmental, social and economic challenges to achieve sustainable development in the Valencian Region's tourism industry.

Recommendations

Considering the importance of an impact culture, the suggestions for the transition towards a circular economy in the Valencian Region's tourism industry include:

- Implementing actions aimed at reducing energy consumption and improving energy efficiency, with special emphasis on sustainable lighting and the use of renewable energies.
- Implementing measures for water-saving and sustainable management of water resources, which is particularly crucial in coastal regions such as the Valencian Region.
- Promoting sustainable production and consumption to reduce its plastic footprint, boosting agroecology and local products, and encouraging sustainable waste management.
- Promoting home automation solutions, electronics and ICT integration, an innovative culture and measures for efficient use of resources and environmental management.
- Taking a holistic view that recognises the links between the different environmental, social and economic aspects of sustainability.
- Counteracting the lack of consumer awareness and rigid legislation that hampers food and plastic waste management.
- Encouraging sustainable consumption and production, and the responsible use of resources and energy-efficient technologies and processes.
- Promoting environmental education and awareness-raising in the tourism industry.
- Implementing measures to address the challenges of energy efficiency, sustainable water-resource management, plastic footprint reduction, the promotion of agroecology and local products, and innovation in the tourism industry.

References

Alonso-Monasterio Fernández, P. (2019). Análisis de la igualdad de género en el sector turístico. El caso de los SICTED de la Comunitat valenciana en 2018. *Papers de Turisme, ISSN 0214-8021, Nº. 62, 2019 (Ejemplar Dedicado a: Igualdad de Género En Turismo), Págs. 1–23, 62*, 1–23. https://dialnet.unirioja.es/servlet/articulo?codigo=6958380&info=resumen&idioma=ENG

Arisi, B. M. (2020). Circular economy from waste to resource: 7Rs innovative practices in Amsterdam. Technologie "die nog in de kinderschoenen staan". *Iluminuras, 21*(55). https://doi.org/10.22456/1984-1191.108061

Capgemini Research Institute. (2021). Circular Economy for a sustainable future. *How organizations can empower consumers and transition to a circular economy.* https://www.capgemini.com/wp-content/uploads/2021/12/Circular-Economy_22122021_v12_web.pdf.

Clark, C. (2014). *The impact investor lessons in leadership and strategy for collaborative capitalism.* In: Emerson, Ben. Thornley (eds) [Book]. Wiley.

de Miguel Molina, B., de Miguel Molina, M., Santamarina Campos, V., & Segarra Oña, M. (2022). *Informe de necesidades del sector turístico para la transición a la economía circular.* https://doi.org/10.5281/ZENODO.6325274.

Einarsson, S., & Sorin, F. (2020). *Circular economy in travel and tourism.* https://circulareconomy.europa.eu/platform/sites/default/files/circular-economy-in-travel-and-tourism.pdf.

Elliot, A. J. (2015, Apr). Color and psychological functioning: A review of theoretical and empirical work. *Frontiers in Psychology, 6,* 368. https://doi.org/10.3389/FPSYG.2015.00368

European Commission. (2020). *Circular economy. Action plan. For a cleaner and more competitive Europe.* https://doi.org/10.2779/05068.

Gobierno de España. (2021). *Componente 14 Plan de modernización y competitividad del sector turístico Contenidos.* https://www.lamoncloa.gob.es/temas/fondos-recuperacion/Documents/16062021-Componente14.pdf.

Heller, E. (2004). In J. Chamorro Mielke (Ed.), *Psicología del color: Cómo actúan los colores sobre los sentimientos y la razón.* Gustavo Gili.

Hervas-Oliver, J. L., Boronat-Moll, C., Sempere-Ripoll, F., & Estelles-Miguel, S. (2018). Plan Sectorial del Plástico, *Plan Estratégico de la Industria Valenciana.* https://portalindustria.gva.es/documents/161328133/164106546/Plan+Sectorial+PLASTICO+2018.pdf/e1763f35-ac0b-4354-94e1-25e1d8d1cac2

Instituto Valenciano de la Competitividad Empresarial. (2021). *Plásticos y cauchos de Comunitat Valenciana.* https://www.ivace.es/index.php/es/component/weblinks/weblink/545-internacional-documentos/548-sectores/411-plasticos?Itemid=100096&task=weblink.go

Instituto Valenciano de Tecnologías Turísticas. (2022). *Estudio sobre el turismo de salud y bienestar en la Comunitat valenciana.* https://invattur.es/uploads/entorno_37/ficheros/63eba235199a31814048112.pdf.

Ministerio para la Transición Ecológica y el Reto Demográfico. (2021). *Plan de recuperación 130 medidas frente al reto demográfico.* https://www.miteco.gob.es/eu/plan_recuperacion_130_medidas_tcm35-528327.pdf

Secretaría Autonómica de Turismo, & Instituto Valenciano de Tecnologías Turísticas. (2020). *Plan estratégico de turismo de la Comunidad Valenciana 2020/2025.* https://www.turismecv.com/wp-content/uploads/2020/07/Plan-Estrategico-de-Turismo-CV-2020-2025-def.pdf

Turisme Comunitat Valenciana. (2022). *Evolución de la actividad turística Comunitat Valenciana.* https://www.turisme.gva.es/opencms/opencms/turisme/es/contents/estadistiquesdeturisme/anuario/turismo/TCV_2021/1_ActividadCV_2021c.

Visit València. (2022). *València, Plan de sostenibilidad turística 2022–2024.* https://fundacion.visitvalencia.com/sites/default/files/media/downloadable-file/files/Plan%20de%20Sostenibilidad%20Tur%C3%ADstica%2022022.pdf

World Travel & Tourism Council, & Harvard T.H. Chan. (2022). *Economy circular.* https://wttc.org/Portals/0/Documents/WTTC-Harvard-LearningInsight-CircularEconomy.pdf

Open Access This chapter is licensed under the terms of the Creative Commons Attribution 4.0 International License (http://creativecommons.org/licenses/by/4.0/), which permits use, sharing, adaptation, distribution and reproduction in any medium or format, as long as you give appropriate credit to the original author(s) and the source, provide a link to the Creative Commons license and indicate if changes were made.

The images or other third party material in this chapter are included in the chapter's Creative Commons license, unless indicated otherwise in a credit line to the material. If material is not included in the chapter's Creative Commons license and your intended use is not permitted by statutory regulation or exceeds the permitted use, you will need to obtain permission directly from the copyright holder.

Chapter 4
Designing a Dynamic Map of Circular Economy in the Tourism Sector of the Valencian Community

Conrado Carrascosa-Lopez, M. Rosario Perello-Marin, and María Ángeles Carabal-Montagud

Introduction

Growing interest is being shown in the circular economy (CE) in areas like strategic management, operations management and technology management. Moreover, governments, industry players and academia are increasingly recognizing and focusing on this concept. Adopting CE principles has become crucial for companies to sustain their competitive advantage (Centobelli et al., 2020; Ghisellini et al., 2016; Pieroni et al., 2019).

CE can be defined as an economic system that aims to eliminate waste and keep resources in continuous use to, thereby, minimize the consumption of raw materials and energy, and the environmental impact (Alcayaga et al., 2019; De Angelis et al., 2023; Murray et al., 2017; Perello-Marin et al., 2022). It is based on closing the loop of the traditional linear "take-make-dispose" model of production and consumption (Blomsma et al., 2019; Pieroni et al., 2019).

Within a CE framework, products and materials are intentionally designed, produced, and utilized to enable reuse, repair, remanufacturing or recycling instead of being disposed of after a single use. The overarching objective of CE is to establish a regenerative system that conserves resources and minimizes waste (Blomsma et al., 2019; Puntillo, 2023). Therefore, in addition to the 3 Rs principles (Reuse, Reduce, Recycle) commonly applied in sustainability, CE also includes

C. Carrascosa-Lopez (✉) · M. R. Perello-Marin
Department of Business Organisation, Universitat Politècnica de València, Valencia, Spain
e-mail: concarlo@upvnet.upv.es; rperell@upvnet.upv.es

M. Á. Carabal-Montagud
Department of Conservation and Restoration of Cultural Assets, Universitat Politècnica de València, Valencia, Spain
e-mail: macamon@crbc.upv.es

© The Author(s) 2024
M. Segarra-Oña et al. (eds.), *Managing the Transition to a Circular Economy*, SpringerBriefs in Business, https://doi.org/10.1007/978-3-031-49689-9_4

efficiency (in energy and resources) and collaborative approaches (Kevin van Langen et al., 2021; Lüdeke-Freund et al., 2019).

Hence the transition to CE requires a systemic shift and collaboration across sectors to redesign products, rethink supply chains and promote sustainable consumption patterns (Puntillo, 2023; Romero-Perdomo et al., 2022). By adopting CE principles, societies can reduce waste generation, conserve natural resources and create economic opportunities, while minimizing environmental impacts (Lüdeke-Freund et al., 2019; Murray et al., 2017; Suchek et al., 2021).

The hospitality sector has significant ecological ramifications and can exert substantial strain on resources because it could involve or be land, water, energy and food. This, in turn, leads to the accumulation of substantial waste and contributes to issues like overcrowding, noise pollution and air pollution (Florido et al., 2019; Rodríguez et al., 2020).

Therefore, applying CE to the tourism sector may provide numerous opportunities to operate more sustainably, reduce costs, enhance reputation and contribute positively to the environment and local communities (Florido et al., 2019; Rodríguez-Antón & Alonso-Almeida, 2019; Rodríguez et al., 2020; Vargas-Sánchez, 2018). By embracing circularity, hotels can align themselves with the growing demand for sustainable tourism and position themselves as leaders in industry (Menegaki, 2018; Pan et al., 2018).

However, this paradigm is not always easy to implement because it requires taking a comprehensive approach that involves various stakeholders, including tourism businesses, governments, local communities, and even tourists themselves (Gamidullaeva et al., 2022; Kevin van Langen et al., 2021; Rodríguez et al., 2020).

One of the main barriers that small- and medium-sized enterprises, especially those in the hospitality sector, find when designing and implementing CE is the considerable difficulty in gaining access to the different stakeholders in the supply chain, which can be aligned with their CE strategies. Not that much information is easily available, and what is accessible is not always easy to use or to understand (Reuter, 2016; Wu, 2020).

Mapping sustainable data emerges as a very effective approach to overcome these obstacles (Reuter, 2016). Mapping data involves organizing and visualizing data in a spatial format, i.e., on a map, to enhance understanding and accessibility (Blomsma et al., 2019; Oymatov et al., 2021; Romero-Perdomo et al., 2022). Thus as mapping makes data easy to use and understand, it empowers users to extract valuable insights, identify trends and make informed decisions (Blomsma et al., 2019; Mies & Gold, 2021; Oymatov et al., 2021; Romero-Perdomo et al., 2022).

By following the model proposed by De Angelis et al. (2023) about circular business models, the Innoecotur team has developed a dynamic map based on the three-pronged strategy framework for circular business models research by focusing on open strategy and developing dynamic capabilities adopted by companies in the hospitality sector.

Methodology to Build the Innoecotur Interactive Map

Dynamic maps provide numerous benefits, especially in tourism. By enhancing their adaptation to circularity, the companies operating in the tourism sector can not only extend their own sectors, but can also contribute to the growth of other industries by implementing comprehensive and transversal policies.

Dynamic maps offer additional benefits because circular companies can locate and connect with one another, which facilitates the analysis and adoption of the best practices implemented by other companies. This allows the creation of synergies and the exchange of knowledge in the CE ecosystem. In particular, the circularity elements that have been identified are interconnected and have implications for various groups of the involved actors (Mies & Gold, 2021).

In the Spanish context, specific circular technology platforms can be found, such as GIEC: "the Interplatform Group for Circular Economy". This group includes 29 Spanish Technology Platforms. Their ongoing initiatives actively promote the implementation of European and Spanish strategies by supporting research and innovation endeavors and by facilitating collaborative projects in national and international programs (GIEC.es, 2022). There are also specific platforms for the packaging value chain from suppliers to recycling (Ecoembes, 2023).

If we look for dynamic maps, two examples can be found in the CE context: one in Spain and another one in Italy. The Spanish case is *EnCircular* (2023) and the Italian one is *Atlante* (2023), but the latter is not specifically from the tourism sector.

When looking in detail on the *EnCircular* website, we find that a CE map of the Valencian Community in Spain is hosted there. This map makes visible the network of organizations and agents committed to evolution toward a "circular" system in the Valencian Region. It aims to be a strategic tool to facilitate alliances and focuses on general sectors, such as industries, education, energy, public administration, technology and telecommunications, among others. It is a general map that also includes good practices of circular companies (Encircular.es, 2023).

In the Italian case, the *Atlante* Map is hosted on the EconomiaCircolare.com website. It was developed by the CDCA, the Documentation Center for Environmental Conflicts, with the support of Erion. *Atlante* is an interactive web platform that researches and reports the experiences of economic realities and associations that are committed to apply CE principles in Italy. The interactive map is intended to act as an awareness-raising, information and documentation tool for all those concerned about striking a balance between economy and ecology, and who wish to orient their consumption in a responsible way (Atlante, 2023).

Yet to date no interactive map has been developed that reflects information on the CE value chain in the tourism sector. For this reason, and by taking into account the numerous benefits that it can bring to the tourism sector in the Valencian Community, the Innoecotur team has developed the first of its kind in this sector.

Conceptual Framework. Identifying the Actors Involved in the Tourism Value Chain in the Valencian Community

Based on the stakeholders in CE in tourism identified by Gamidullaeva et al. (Gamidullaeva et al., 2022), namely tourism businesses, governments, local communities and tourists, an analysis of the tourism CE in the Valencia Region was performed, which ranged from service providers to end consumers. This analysis comes as a dynamic map to make it easier for all the aforementioned stakeholders to locate and connect with one another to, thus, facilitate the analysis and adoption of the best practices implemented by other companies.

The main goal is to map the CE tourism value chain to have a stronger impact on local communities and tourists themselves. It is important to highlight that local governments not only have the power to influence the value proposition of their region's tourism industry through regulations, but can also directly experience tourism ecosystem outcomes. They bear the responsibility of managing the socioeconomic development and efficiency of the tourism sector by ensuring its positive impacts on the region, while minimizing any potential negative effects (Gamidullaeva et al., 2022; Marjamaa et al., 2021; Menegaki, 2018).

Therefore, when considering all these aspects, the key actors included in the present project are the following (see Fig. 4.1):

1. Accommodation providers: this category includes hotels, resorts, guesthouses, bed & breakfasts, and other types of accommodation establishments.
2. Restaurants and food service providers: the region is known for its culinary offerings, and restaurants, cafes, bars, and other food service providers play a vital role in serving tourists local cuisine.

Fig. 4.1 Conceptual framework on the Innoecotur map of CE. Source: the Innoecotur project

3. Different suppliers that attend to the necessities of the above-mentioned main actors, such as food and drinks, energy, water, textiles, construction, consultancy, hygiene, lighting, furniture, paper, transport, logistics, among others.
4. Local policies. All the above-mentioned actors are affected by the local policies applied in the region.

Aspects to Analyze the Environmental Impact of These Actors and Their Contribution to Circular Economy

In order to identify actors' contribution to CE, an evaluation of their practices, operations and initiatives, aligned with CE principles, was made. In doing so, different methods could have been considered, but the Innoecotur team decided to focus on certifications and sustainable seals by paying attention to Sustainable Development Goals (United Nations, 2023).

For the search and analysis of the most appropriate certificates and seals, the Innoecotur team assessed the central aspects of CE, as stated above: Product design for durability and reuse; Reduce, refurbish, remanufacture; Recycle and recover materials; Renewable energy and resource efficiency. These aspects (Segarra Oña et al., 2023) were grouped into four categories (see Fig. 4.2):

Energy: reduce, resource efficiency.
Water: reduce, resource efficiency.
Food & drinks: waste reduction, resource efficiency.
Materials & waste: durability & reuse; reduce; recycle & recover, resource efficiency.

The transversal policies that affect all the above categories were grouped into these categories, where we can find aspects like technology, other suppliers, consultancy, mobility or training and employment.

The process followed to build the interactive map was done by identifying the main actors, together with the principal certifications and seals, related to the covered aspects.

To ensure the credibility of the companies featured on the map, and to ascertain their actual contributions and the specific areas they impact, it is crucial to demonstrate and verify all the aforementioned aspects in relation to CE.

This is why certifications and seals are especially relevant because they are a way to demonstrate that the companies included on the map have followed a verification process about the CE aspects they claim, and there is formal evidence to demonstrate their CE involvement.

A worldwide search for certifications was made. Only the certifications and seal that deal with the aspects included within this conceptual framework were considered. Table 4.1 lists the certifications included on the map, together with the section of the conceptual map that they mainly contribute to.

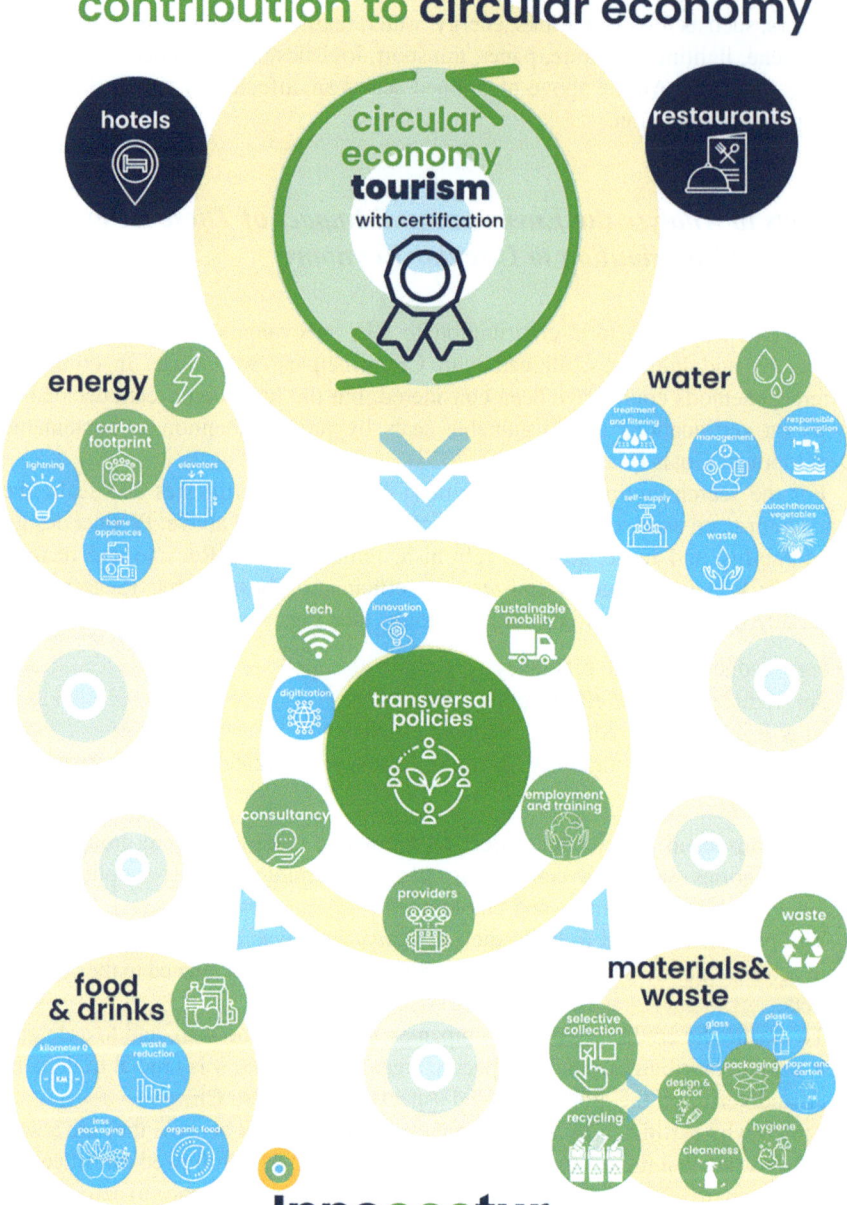

Fig. 4.2 Contribution to the circular economy of the agents involved in tourism. Source: the Innoecotur project

Table 4.1 Contribution to the circular economy of certifications

Certification	Contribution	Certification	Contribution	Certification	Contribution
Bcorp	Energy	Booking viajes sostenibles	Tourism	ISO 14046	Water
Carbon proof	Energy	Green Globe	Tourism		
Coop electricas	Energy	Green Key	Tourism	Certification	Contribution
Huella Carbono Ministerio	Energy	Hilton Lightstay	Tourism	Biosphere	Transversal
		Hostelling International Qualit	Tourism	Blueangel	Transversal
Certification	Contribution	S ICTE	Tourism	Cisne Blanco	Transversal
Acene	Food & Drinks	Travelife	Tourism	Cradle to cradle	Transversal
ASC Aquaculture Stewardship Co	Food & Drinks			Earthcheck	Transversal
Bienestar animal	Food & Drinks	Certification	Contribution	Ecolabel	Transversal
Bio vida Sana	Food & Drinks	Breem	Materials	EMAS	Transversal
CAECV	Food & Drinks	Global Recycled Standard	Materials	Madera Justa CV	Transversal
CosmeBio	Food & Drinks	GOTS textil	Materials	Preferred by nature	Transversal
Demeter	Food & Drinks			REAS	Transversal
Ecocert	Food & Drinks			Sannas	Transversal
Fair trade	Food & Drinks				

Source: the Innoecotur project

Identify the Sustainable and Circular Practices That Are Being Implemented into the Tourism Industry in the Valencian Community with a Dynamic Map

As previously mentioned, information on a map makes it easier to present and to search in a very visual intuitive way. This is why the Innoecotur team used this format to display information.

The map is constantly updated and the last version can be viewed on this website: https://innoecotur.webs.upv.es/mapa/. It is an open source that is available to any user (see Fig. 4.3):

This map is labeled as a dynamic map because it is not fixed, but is updated at any time when companies in the CE tourism cluster join it. It is designed to be open. It is constantly updated with the data from all those companies that adapt and introduce aspects of circularity into their day-to-day operations. The purpose is to increase their visibility and acknowledgement by as many companies as possible to serve as a reference.

This map can facilitate the search for suppliers with demonstrated concern and circular mentality to increase the circularity of the entire ecosystem.

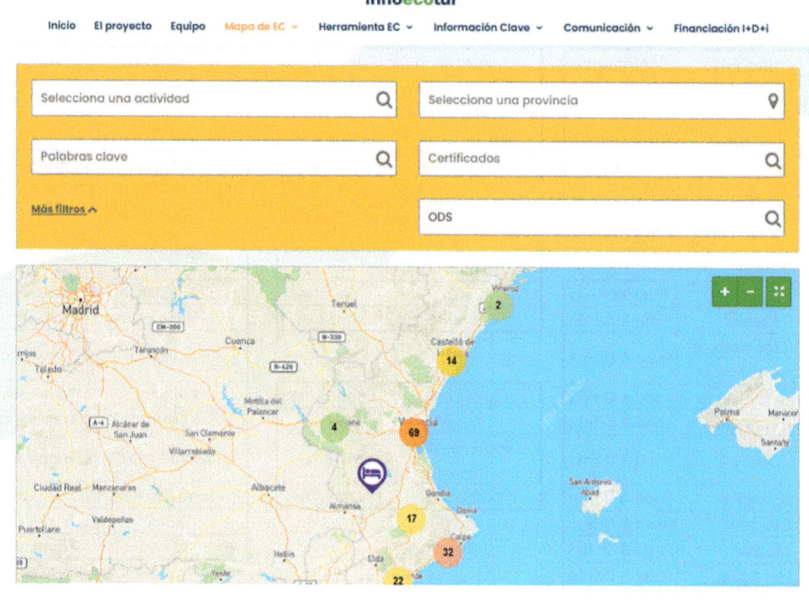

Fig. 4.3 Interactive Circularity Map Search Engine aspect view and example. Source: the Innoecotur project

Data value filtering, which is intrinsic to the visualization process, was employed to facilitate shifting the focus between different data subsets to analyze specific categories of values (Heer et al., 2012).

The map incorporates all the companies with any of the certifications mentioned in the tables and located in the Valencian Community. They were grouped according to activities, certificates and geographical area or province. Searches can be made for activities, specific certificates or geographical areas.

Conclusions

After developing this dynamic map, it can be stated that it is not only a handy tool that makes visible the companies involved in circularity in tourism sector, but it is also a tool that can act as a lever for change to empower and motivate increasingly more companies to adopt actions that benefit their circularity and sustainability.

This map also shows the wide variety of certifications and seals that deal with CE. They include details, such as environmental aspects, energy efficiency, waste reduction, resources optimization, proximity of resources, among others.

As a preliminary conclusion of this analysis of the different CE-related certifications and seals, lots of duplicities in their approaches appear. Many cover the same concepts, but with minor variations. However, we have yet to find any that broadly covers circular behavior. We believe that this fact hinders the visibility of CE and, thus, makes it difficult for companies to adopt it.

During the process of providing this map with content, the team was asked for information by many companies that wished to adopt CE practices. They asked us for recommendations on which certification could meet these concerns, or one that would cover as many CE requirements as possible. The answer to the question is complex; there are many, but they are all incomplete. Some are complementary and others overlap part of the scope.

As a general conclusion of this work, we stress the need to develop certification that clearly covers all CE aspects as a whole.

References

Alcayaga, A., Wiener, M., & Hansen, E. G. (2019). Towards a framework of smart-circular systems: An integrative literature review. *Journal of Cleaner Production, 221*, 622–634. https://doi.org/10.1016/j.jclepro.2019.02.085

Atlante. (2023). *Economia Circolare.com*. https://economiacircolare.com/atlante/

Blomsma, F., Pieroni, M., Kravchenko, M., Pigosso, D. C. A., Hildenbrand, J., Kristinsdottir, A. R., Kristoffersen, E., Shabazi, S., Nielsen, K. D., Jönbrink, A. K., Li, J., Wiik, C., & McAloone, T. C. (2019). Developing a circular strategies framework for manufacturing companies to support circular economy-oriented innovation. *Journal of Cleaner Production, 241*, 118271. https://doi.org/10.1016/J.JCLEPRO.2019.118271

Centobelli, P., Cerchione, R., Chiaroni, D., Del Vecchio, P., & Urbinati, A. (2020). Designing business models in circular economy: A systematic literature review and research agenda. *Business Strategy and the Environment, 29*(4), 1734–1749. https://doi.org/10.1002/BSE.2466

De Angelis, R., Morgan, R., & De Luca, L. M. (2023). Open strategy and dynamic capabilities: A framework for circular economy business models research. *Business Strategy and the Environment., 32*, 4861–4873. https://doi.org/10.1002/bse.3397

Ecoembes. (2023). *Ecoembes*. https://www.ecoembes.com/es

Encircular.es. (2023). *Mapa EnCircular*. https://encircular.es

Florido, C., Jacob, M., & Payeras, M. (2019). How to carry out the transition towards a more circular tourist activity in the hotel sector. The role of innovation. *Administrative Sciences, 9*(2). https://doi.org/10.3390/ADMSCI9020047

Gamidullaeva, L., Vasin, S., Tolstykh, T., & Zinchenko, S. (2022). Approach to regional tourism potential assessment in view of cross-sectoral ecosystem development. *Sustainability (Switzerland), 14*(22). https://doi.org/10.3390/su142215476

Ghisellini, P., Cialani, C., & Ulgiati, S. (2016). A review on circular economy: The expected transition to a balanced interplay of environmental and economic systems. *Journal of Cleaner Production, 114*, 11–32. https://doi.org/10.1016/J.JCLEPRO.2015.09.007

GIEC.es. (2022). *Grupo Interplataformas de Economía Circular©*.

Heer, J., Shneiderman, B., & Park, C. (2012). A taxonomy of tools that support the fluent and flexible use of visualizations. *Interactive Dynamics for Visual Analysis, 10*, 1–26. http://queue.acm.org/detail.cfm?id=2146416

Kevin van Langen, S., Vassillo, C., Ghisellini, P., Restaino, D., Passaro, R., & Ulgiati, S. (2021). Promoting circular economy transition: A study about perceptions and awareness by different stakeholders groups. *Journal of Cleaner Production, 316*, 128166. https://doi.org/10.1016/J.JCLEPRO.2021.128166

Lüdeke-Freund, F., Gold, S., & Bocken, N. M. P. (2019). A review and typology of circular economy business model patterns. *Journal of Industrial Ecology, 23*(1), 36–61. https://doi.org/10.1111/JIEC.12763

Marjamaa, M., Salminen, H., Kujala, J., Tapaninaho, R., & Heikkinen, A. (2021). A sustainable circular economy: exploring stakeholder interests in Finland. *South Asian Journal of Business and Management Cases, 10*(1), 50–62. https://doi.org/10.1177/2277977921991914

Menegaki, A. N. (2018). Economic aspects of cyclical implementation in Greek sustainable hospitality. *International Journal of Tourism Policy, 8*(4), 271–302. https://doi.org/10.1504/IJTP.2018.098896

Mies, A., & Gold, S. (2021). Mapping the social dimension of the circular economy. *Journal of Cleaner Production, 321*. https://doi.org/10.1016/j.jclepro.2021.128960

Murray, A., Skene, K., & Haynes, K. (2017). The circular economy: An interdisciplinary exploration of the concept and application in a global context. *Journal of Business Ethics, 140*(3), 369–380. https://doi.org/10.1007/S10551-015-2693-2

Oymatov, R. K., Mamatkulov, Z. J., Reimov, M. P., Makhsudov, R. I., & Jaksibaev, R. N. (2021). Methodology development for creating agricultural interactive maps. *IOP Conference Series: Earth and Environmental Science, 868*(1). https://doi.org/10.1088/1755-1315/868/1/012074

Pan, S. Y., Gao, M., Kim, H., Shah, K. J., Pei, S. L., & Chiang, P. C. (2018). Advances and challenges in sustainable tourism toward a green economy. *Science of the Total Environment, 635*, 452–469. https://doi.org/10.1016/J.SCITOTENV.2018.04.134

Perello-Marin, M. R., Carrascosa-López, C., De Miguel Molina, M., & Mas Gil, M. A. (2022, January 10). *Lost in the forest of circular economy certificates in tourism sector*. 4th International Conference Business Meets Technology. https://doi.org/10.4995/bmt2022.2022.15640.

Pieroni, M. P. P., Mcaloone, T. C., & Pigosso, D. C. A. (2019). Business model innovation for circular economy and sustainability: A review of approaches. *Journal of Cleaner Production, 215*, 198–216. https://doi.org/10.1016/j.jclepro.2019.01.036

Puntillo, P. (2023). Circular economy business models: Towards achieving sustainable development goals in the waste management sector—Empirical evidence and theoretical implications. *Corporate Social Responsibility and Environmental Management, 30*(2), 941–954. https://doi.org/10.1002/CSR.2398

Reuter, M. A. (2016). Digitalizing the circular economy: Circular economy engineering defined by the metallurgical internet of things. *Metallurgical and Materials Transactions B: Process Metallurgy and Materials Processing Science, 47*(6). https://doi.org/10.1007/s11663-016-0735-5

Rodríguez, C., Florido, C., & Jacob, M. (2020). Circular economy contributions to the tourism sector: A critical literature review. *Sustainability, 12*(11), 4338. https://doi.org/10.3390/SU12114338

Rodríguez-Antón, J. M., & Alonso-Almeida, M. D. M. (2019). The circular economy strategy in hospitality: A multicase approach. *Sustainability (Switzerland), 11*(20). https://doi.org/10.3390/SU11205665

Romero-Perdomo, F., Carvajalino-Umaña, J. D., Moreno-Gallego, J. L., Ardila, N., & González-Curbelo, M. Á. (2022). Research trends on climate change and circular economy from a knowledge mapping perspective. *Sustainability (Switzerland), 14*(1). https://doi.org/10.3390/su14010521

Segarra Oña, M. V., Peiró Signes, A., & Pérez Herrero, M. (2023). *Herramienta básica de autoevaluación de circularidad en hoteles Innoecotur*. https://innoecotur.webs.upv.es/herramienta-ec/

Suchek, N., Fernandes, C. I., Kraus, S., Filser, M., & Sjögrén, H. (2021). Innovation and the circular economy: A systematic literature review. *Business Strategy and the Environment, 30*(8), 3686–3702. https://doi.org/10.1002/BSE.2834

United Nations. (2023). *Sustainable development goals*. https://sdgs.un.org/goals

Vargas-Sánchez, A. (2018). The unavoidable disruption of the circular economy in tourism. *Worldwide Hospitality and Tourism Themes, 10*(6), 652–661. https://doi.org/10.1108/WHATT-08-2018-0056

Wu, Q. (2020). Greemap: A Python package for interactive mapping with Google Earth Engine. *Journal of Open Source Software, 5*(51), 2305. https://doi.org/10.21105/joss.02305

Open Access This chapter is licensed under the terms of the Creative Commons Attribution 4.0 International License (http://creativecommons.org/licenses/by/4.0/), which permits use, sharing, adaptation, distribution and reproduction in any medium or format, as long as you give appropriate credit to the original author(s) and the source, provide a link to the Creative Commons license and indicate if changes were made.

The images or other third party material in this chapter are included in the chapter's Creative Commons license, unless indicated otherwise in a credit line to the material. If material is not included in the chapter's Creative Commons license and your intended use is not permitted by statutory regulation or exceeds the permitted use, you will need to obtain permission directly from the copyright holder.

Part II
Good Practices

Chapter 5
Development of a Model for the Application of the Circular Economy in Hotels and Restaurants Through the 'Customer Journey Map'

Joaquín Sánchez-Planelles, Yolanda Trujillo-Adriá, and Gabriela Ribes-Giner

Introduction

Sustainability is a concept that is being extended to more and more areas of our society, being very present in the business sphere through the so-called corporate sustainability. An increasing number of companies are taking the decision to apply sustainable practices in different areas of their organization, either to face regulatory requirements or simply to satisfy the pressures and demands of their customers.

However, the ability to determine which sustainable practices to apply in order to generate the highest return for companies still seems complicated. This is due to the fact that different variables intervene—for example, the type of sector in which the organizations operate and the level of development of sustainable practices applied by competitors.

Environmental practices and, in particular, the circular economy, are adopted both at strategic and operational level through the so-called environmental, social and governance (ESG) criteria (Eccles et al., 2014). However, knowing what practices to apply or how to integrate the circular economy into the business model is still a complex process for organizations (Galbreath, 2009; Hahn, 2013). This is due to the fact that this decision-making process involves taking into account numerous factors if referring to a specific study—for example, the sector in which the company competes, the demands of stakeholders, internal processes and structures, etc. (Baumgartner, 2014). In fact, there are authors (Engert et al., 2016) who argue that developing a sustainable strategy that determines which sustainable practices to

---

J. Sánchez-Planelles (✉) · Y. Trujillo-Adriá · G. Ribes-Giner
Department of Business Organisation, Universitat Politècnica de València, Valencia, Spain
e-mail: joasanpl@ade.upv.es; yotruad@etsid.upv.es; gabrigi@omp.upv.es

implement requires an ad hoc study that includes the circumstances associated with each organization.

Purpose

In order to help companies in the Valencian tourism sector, especially hotel and restaurant companies, a project called 'Innoecotur' has been designed in which the main Valencian universities participate. This present study is the result of the first phases of the project's work.

The aim of this work is to provide a model that can assist hotel and catering companies in their transition to a circular business model. Specifically, it shows the main circular practices that can be applied along the different points of contact that customers have with accommodation and catering services.

State of the Art

The circular economy (CE) is defined as the system that replaces the 'end-of-life' concept of the existing linear model, promoting the reduction, reuse, recycling and recovery of materials (Kirchherr et al., 2017). In other words, it is an opportunity to create value in a way that benefits society, business and the environment while contributing to the fulfilment of the Sustainable Development Goals (SDGs). According to the Ellen MacArthur Foundation (EMF), the widely used framework in CE research is based on three design-driven principles: eliminating waste and pollution; making products and materials circular; and lastly, regenerating nature (EMF, 2021). The Capgemini Research Institute (2021) shows a series of actions (referred to as the 7 'Rs') that can help organizations assess their current impact: reduce; reuse; redesign; repair/refurbish; restore/remanufacture; return/recover; recycle.

CE-related practices can be tailored to each type of company. The tourism industry is a vast and complex industry that encompasses a variety of sectors and connects with many other industries and value chains (Font & Lynes, 2018; Einarsson & Sorin, 2020; Rosato et al., 2021). In this case, this paper offers a study on how the tourism sector could integrate these types of practices throughout its value chain.

In this case, hotels and restaurants evaluate the journey that their customers follow during the interaction with their services to identify where in the value chain it is possible to implement more sustainable practices, either directly or through their suppliers.

Methodology

A qualitative method is used to explore these circular activities. After reviewing the literature, some reports from companies and organizations and the experiences of some companies through focus groups (De Miguel-Molina et al., 2022), the method selected to obtain information on the needs of companies in the application of the circular economy in hotels and restaurants is the focus group. Through this methodology, the integration of the companies is pursued from the beginning of the project. It is an information-gathering technique that seeks to find out what the participants think/feel about an idea (Krueger, 2015).

Service process models have been applied to follow all the steps that could contribute to a circular model (Kirchherr et al., 2017; Geissdoerfer et al., 2017). To represent the flow of these service processes, we have used the customer journey map (Tueanrat et al., 2021; Lemon & Verhoef, 2016) and infographics (Gareau et al., 2015) for a more visual representation.

Results

Tables 5.1, 5.2, 5.3, 5.4, 5.5, 5.6, 5.7, 5.8, 5.9 represent the main points of interaction between customers and hotel and restaurant services. Additionally, the potential circular practices that could be developed in each process are shown and, finally, what type of suppliers can help companies to undertake these circular practices.

Table 5.1 Booking process for hotels and restaurants

Process	Elements with which it interacts	Circular practices	Supplier activity
Search, comparison and reservation of hotel/restaurant	Website, search engines (Google, Trivago, Booking.com, etc.)/Reception (telephone reservations, email)/Reservation system	Use of environmentally conscious websites (e.g. compensation of the carbon footprint through planting of tree species).	Specific software for accommodation and reservations
Payment of the reservation	Online payment through payment platforms/Payment at reception (by card or cash)/Bank transfer	Reduction of printed documents/Digital discount vouchers	Digital payment platforms

Source: Authors

Table 5.2 Processes associated with the trip (hotels/restaurants)

Process	Elements with which it interacts	Circular practices	Supplier activity
Transfer to the hotel/restaurant	Private vehicle, rented vehicle or public transport (metro, bus, taxi, etc.)	Proposal of sustainable transport through the corporate pages of businesses (e.g. train, electric vehicles, etc.)/Agreements with environmentally responsible transport companies/Suggest customersuse public transport	Mobility and sustainable transport
Parking	Private vehicle or rented vehicle and parking	Existence of electric chargers in the hotel car park/Bicycles to be rented by hotel guests	Electric vehicle chargers, bicycle manufacturer

Source: Authors

Table 5.3 Check-in process at the hotel

Process	Elements with which it interacts	Circular practices	Supplier activity
Wait in line to be attended or 'self-service check-in"	Hotel reception	Use of digital systems to avoid the use of paper	Specific software for accommodation and reservations
Data collection (e.g. DNI)	Hotel reception	Use of digital systems to avoid the use of paper	Specific software for accommodation and reservations
Delivery of the room key in a cardboard envelope	Digital keys, physical keys, etc.	Reduction of the use of cardboard, envelopes, etc. If used, they are made of recycled cardboard/Use of QR codes	Door-opening systems using fingerprints, use of recycled plastic, recycled paper or cardboard
Delivery of paper with access codes for Wi-Fi	Physical documents	Reduction of the use of cardboard, envelopes, etc. If used, they are made of recycled cardboard/Use of QR codes	QR codes, recycled paper or cardboard
Delivery of welcome gifts/'amenities'	E.g. bottle of water	Sustainable gift giving/Promote the use of products with lower environmental impact/Eliminate single-use plastics	'Amenities' with less environmental impact

Source: Authors

Discussion

This work is part of the first phases of development of the Innoecotur project, so that soon the thesis shown here will be validated. In addition, part of the research is developed through case studies hand in hand with actors in the tourism sector. Therefore, this model has been proposed to hotels and restaurants, especially small

5 Development of a Model for...

Table 5.4 Access process (hotels and restaurants)

Process	Elements with which it interacts	Circular practices	Supplier activity
Use of the elevator	Lift	Efficient electrical and lighting systems/Renewable energy	Renewable energy marketer/LED lighting systems
Use of stairs	Stairs	Efficient electrical and lighting systems/Renewable energy	Renewable energy marketer/LED lighting systems

Source: Authors

and medium-sized enterprises (SMEs), as well as to different suppliers that are part of their supply chain.

On the other hand, two infographics are shown (Figs. 5.1 and 5.2) in order to visualize the processes in a clear and simple way to show how a circular model could be developed for hotel and restaurant companies. Small and medium-sized hotels and restaurants in the Valencian community have little access to information on this type of practice, so this work allows them to provide them with a clear and understandable model that can help them, as well as contribute to the literature on CE in the tourism sector.

Limitations

The qualitative data obtained through the realization of three focus groups with stakeholders of the tourism sector are from the Valencian Community in Spain, specifically, so the results of this work may not be representative in all geographical areas. The results of this work will be used to develop a platform where hotels and restaurants can find all kinds of suppliers of a variety of solutions with different approaches on sustainable practices to facilitate the implementation of circular actions in the Valencian Community.

Table 5.5 Processes associated with the use of the room (hotels)

Process	Elements with which it interacts	Circular practices	Supplier activity
Opening suitcases	Furniture—where to locate suitcases, travel bags, etc.	Furniture made using circular practices (e.g. furniture made from plastic bottle waste)	Recycled and reusable furniture
Placing clothes in closets	Hangers	Hangers made by circular practices	Recycled and reusable plastic
Know the services of the hotel	Book or catalogue with hotel services/Bar and restaurant menu	Digital system. In case of being physical, be made with paper and recycled materials	QR codes, recycled and reusable paper
Request for food and drink through room service	Hotel kitchen	Avoid single-use plastics/Use sustainable and reusable utensils and tableware	Cutlery and utensils made of recycled and reusable materials
Personal hygiene	Sanitary hot water, shampoo, soap, lotions, towels, toilet paper, sanitary bags, etc.	Hygiene products developed through sustainable practices/Avoid single-use plastics/Water consumption reduction systems (e.g. aerators in taps, pushbuttons in bathroom taps, small toilet cisterns)/Awareness of water consumption/Products purchased in bulk (minimum packaging)/Eco-friendly textiles	Water-reuse systems, renewable energy, green chemicals, reusable packaging, green textiles
Hair drying	Guest-owned hairdryer or hairdryer offered by the hotel	Installation of dryers with energy efficiency label	Energy-efficient appliances
Rest	Bed, sheets, pillows, cushions, mattress, etc.	Textile materials made using circular practices	Ecological textiles
Temperature control	Air-conditioning and heating systems	Equipment with energy-efficiency certification/Awareness of energy consumption/Use of home automation or other technologies to avoid use when there is no one in the room or windows are open/Sustainable energy sources	Energy-efficient cooling and heating systems, use of renewable energy
Use of the minibar	Minibar fridge and the products inside	Energy-efficient refrigerators/Sale of products in the minibar that come from sustainable brands/Recyclable packaging/	Renewable energy, organic food and beverages, reusable packaging

(continued)

Table 5.5 (continued)

Process	Elements with which it interacts	Circular practices	Supplier activity
		Filtered water instead of bottles	
Room cleaning	Chemicals used for cleaning, gloves, garbage bags, refills of hygiene products, etc.	Sensitize customers to reuse towels and sheets/ Use of cleaning products made with sustainable criteria (e.g. easily recyclable garbage bags)/ Cleaning chemicals with low impact on the environment/Reusable or returnable packaging/ Waste separation containers	Water-reuse systems, green chemicals, recycled and reusable packaging
Whites wash	Laundry inside the hotel or outsourced to a supplier	Machines with energy certification/Minimum water consumption/Supplier with sustainable standards/Low environmental impact detergent/ Reuse of washing water	Water-reuse systems, use of renewable energy, use of green chemicals

Source: Authors

Table 5.6 Processes associated with table food service (restaurants)

Process	Elements with which it interacts	Circular practices	Supplier activity
Menu consultation	Restaurant menu	Digital system. In case of being physical, be made with paper and recycled materials	Recycled paper
Food & Beverage order	Notepad or note-taking system	Use of digital systems instead of paper	Recycled and reusable paper
Use of the bathroom	Sanitary water, soap, paper, hand dryer, etc.	Hygiene products developed through sustainable practices/ Avoid single-use plastics/ Water consumption reduction systems (e.g. aerators in taps, pushbuttons in bathroom taps, small toilet cisterns)/Water consumption awareness/Products purchased in bulk (minimum packaging)	Water-reuse systems, renewable energy, green chemicals, reusable packaging
Comfort temperature	Air-conditioning and heating systems	Equipment with energy-efficiency certification/ Awareness of energy consumption/Use of home automation or other technologies to avoid use when there is no one in the room or windows are open/Sustainable energy sources	Energy-efficient cooling and heating systems, use of renewable energy
Cleaning	Chemicals, gloves, garbage bags, spare cleaning products	Use of cleaning products made with sustainable criteria (e.g. easily recyclable garbage bags)/Cleaning chemicals with low impact on the environment/Reusable or returnable packaging/Containers to separate waste	Water-reuse systems, renewable energy, green chemicals, reusable packaging, green textiles
Menu consultation	Restaurant menu	Digital system. In case of being physical, be made with paper and recycled materials	Recycled paper
Food & Beverage order	Notepad or note-taking system	Use of digital systems instead of paper	Recycled and reusable paper
Use of the bathroom	Sanitary water, soap, paper, hand dryer, etc.	Hygiene products developed through sustainable practices/ Avoid single-use plastics/ Water consumption reduction systems (e.g. aerators in taps, pushbuttons in bathroom taps, small toilet cisterns)/Water consumption awareness/	Water-reuse systems, renewable energy, green chemicals, reusable packaging

(continued)

Table 5.6 (continued)

Process	Elements with which it interacts	Circular practices	Supplier activity
		Products purchased in bulk (minimum packaging)	
Comfort temperature	Air-conditioning and heating systems	Equipment with energy-efficiency certification/ Awareness of energy consumption/Use of home automation or other technologies to avoid use when there is no one in the room or windows are open/Sustainable energy sources	Energy-efficient cooling and heating systems, use of renewable energy
Cleaning	Chemicals, gloves, garbage bags, spare cleaning products	Use of cleaning products made with sustainable criteria (e.g. easily recyclable garbage bags)/Cleaning chemicals with low impact on the environment/Reusable or returnable packaging/Containers to separate waste	Water-reuse systems, renewable energy, green chemicals, reusable packaging, green textiles

Source: Authors

Table 5.7 Check out processes (hotels)

Process	Elements with which it interacts	Circular practices	Supplier activity
Wait at reception to check out	Reception	Digital system that allows you to check out autonomously	Specific software for hotels
Delivery of the key together with the cardboard envelope	Reception	Reduction of the use of cardboard envelopes. If used, they must be made using recycled cardboard	Opening systems using fingerprint, plastics and recycled paper/cardboard
Delivery of the invoice to the customer	Reception	Sending the invoice online/Use of recycled paper	Specific software for hotels, recycled paper
Delivery to the customer of the satisfaction survey	Reception	Survey that can be filled out digitally/Using digital tablets to fill it out/Sending the survey to the customer through the mail	QR codes, recycled paper, specific software for hotels
Wait at reception to check out	Reception	Digital system that allows you to check out autonomously	Specific software for hotels

Source: Authors

Table 5.8 Other

Process	Elements with which it interacts	Circular practices	Supplier activity
Gymnasium	Gym room, exercise machines, sports equipment (e.g. weights)	Energy-efficient machines/LED lighting system/Towels made with sustainable textiles	Specific software for hotels
Pool/SPA	Changing rooms, bathrooms, swimming pool, jacuzzi, etc.	Environmentally friendly chemicals/Sustainable heating systems (efficient boilers, biomass boilers, solar panels for domestic hot water, LED lighting, etc.)	Water-reuse or -saving systems/Use of renewable energies/Ecological chemicals
Bicycles	Bicycles and where bicycles are located	Promote their use/Offer parking for own and guest bicycles/Manufacture of bicycles using circular practices	Bicycle producers

Source: Authors

Table 5.9 Cross-cutting actions to hotels and restaurants

Process	Elements with which it interacts	Circular practices	Supplier activity
Construction	Building materials	Durable building materials/Techniques that promote energy efficiency (e.g. insulation, ventilation, etc.)/Building certification	Sustainable construction
Suppliers	Supplier management	Promoting the procurement of goods and services from suppliers that have environmental certificates/Prioritizing local suppliers/Collaborating to improve supplier practices	Supplier evaluation and risk analysis systems
Water	Technology for the reduction of water consumption and its reuse	Wastewater or rainwater reuse systems for irrigation systems/Water consumption reduction (e.g. aerators in taps, pushbuttons in bathroom taps, small toilet cisterns)	Water-reduction and reuse systems
Customer and employee awareness	Communication elements	Posters/Conferences & events/Sustainability in advertising/Communicating sustainable practices/Integrating sustainability into business strategy	Environmental consulting
Waste management	Containers/Waste collection/Waste management	Waste separation containers/Return of returnable materials to suppliers/Composting bin/Manage the end of life of utensils and furniture	Waste managers

(continued)

Table 5.9 (continued)

Process	Elements with which it interacts	Circular practices	Supplier activity
Appearance of hotels and restaurants	Furniture and decoration	Recyclable materials/Sustainable manufacturing	Furniture manufactured by circular practices
Textile recycling	Textiles	Recycling of linen (sheets, towels) and uniforms	Textile material manufactured by circular practices

Source: Authors

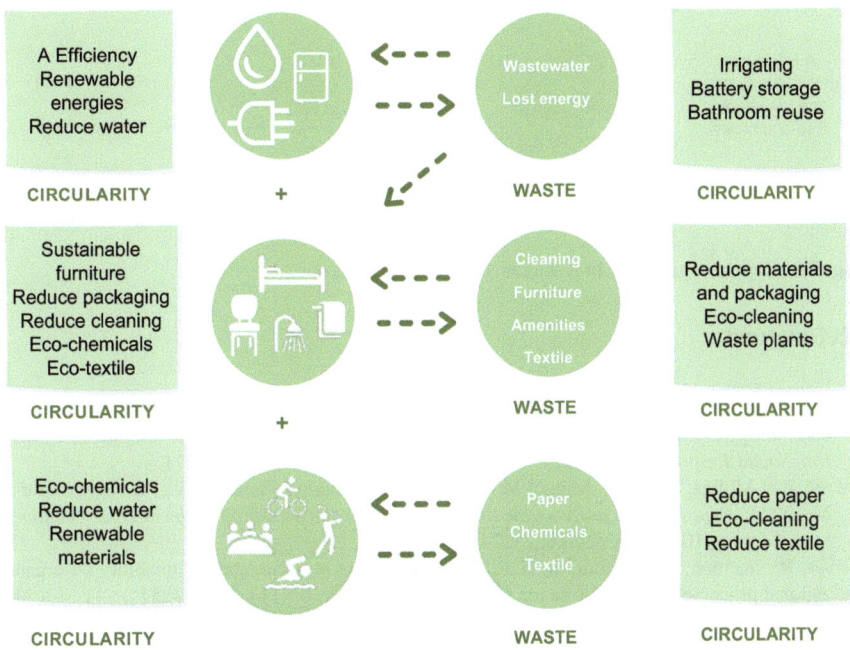

Fig. 5.1 Circular operations in hotels. Source: Authors

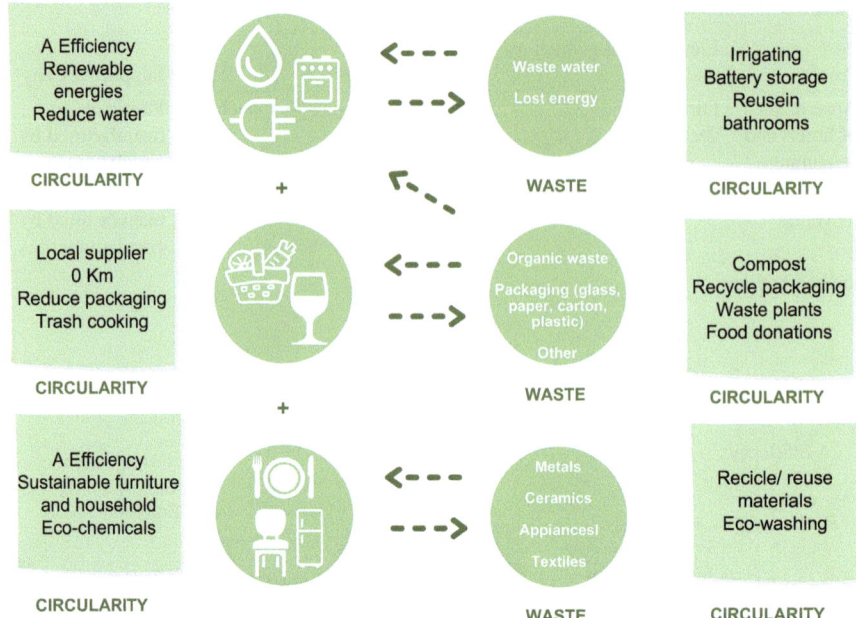

Fig. 5.2 Circular operations in restaurants. Source: Authors

References

Baumgartner, J. R. (2014). Managing corporate sustainability and CSR: A conceptual framework combining values, strategies and instruments contributing to sustainable development. *Corporate Social Responsibility and Environmental Management, 21*(5), 258–271.
De Miguel-Molina, B., De Miguel-Molina, M., Santamarina-Campos, V., & Segarra-Oña, M. (2022). *Report on the needs of the tourism sector for the transition to the circular economy*. https://doi.org/10.5281/zenodo.6325274
Eccles, R., Ioannou, I., & Serafeim, G. (2014). The impact of corporate sustainability on organizational processes and performance. *Management Science, 60*(11), 2835–2857.
Einarsson S., & Sorin, F. (2020). *Circular economy in travel and tourism: A conceptual framework for a sustainable, resilient and future-proof industry transition, CE360 Alliance*.
Ellen MacArthur Foundation – EMF. (2021). *Completing the picture: How the circular economy tackles climate change*. https://emf.thirdlight.com/link/w750u7vysuy1-5a5i6n/@/preview/1?oMolina, D. B. M. (2022, March 3).
Engert, S., Rauter, R., & Baumgartner, R. J. (2016). Exploring the integration of corporate sustainability into strategic management: A literature review. *Journal of Cleaner Production, 112*, 2833–2850.
Font, X., & Lynes, J. (2018). Corporate social responsibility in tourism and hospitality. *Journal of Sustainable Tourism, 26*(7), 1027–1042.
Galbreath, J. (2009). Building corporate social responsibility into strategy. *European Business Review, 21*(1), 109–127.
Gareau, M., Keegan, R., & Wang, L. (2015). An exploration of the effectiveness of infographics in contrast to text documents for visualizing census data: What works? In S. Yamamoto (Ed.), *Human interface and the management of information. Information and knowledge design. HIMI*

2015. Lecture Notes in Computer Science, 9172. Springer. https://doi.org/10.1007/978-3-319-20612-7_16

Geissdoerfer, M., Savaget, P., Bocken, N., & Hultink, E. (2017). The circular economy: A new sustainability paradigm? *Journal of Cleaner Production, 143,* 757–768.

Hahn, R. (2013). ISO 26000 and the standardization of strategic management processes for sustainability and corporate social responsibility. *Business Strategy and the Environment, 22*(7), 442–455.

Kirchherr, J., Reike, D., & Hekkert, M. (2017). Conceptualizing the circular economy: An analysis of 114 definitions. *Resources, Conservation and Recycling, 127,* 221–232.

Krueger, P. (2015). Corporate goodness and shareholder wealth. *Journal of Financial Economics, 115*(2), 304–329.

Lemon, K. N., & Verhoef, P. C. (2016). Understanding customer experience throughout the customer journey. *Journal of Marketing, 80*(6), 69–96. https://doi.org/10.1509/jm.15.0420

Rosato, P. F., Caputo, A., Valente, D., & Pizzi, S. (2021) 2030 Agenda and sustainable business models in tourism: A bibliometric analysis. *Ecological Indicators,* 121pp.

Tueanrat, Y., Papagiannidis, S., & Alamanos, E. (2021). Going on a journey: A review of the customer. *Journal of Business Research, 125,* 336–353.

Open Access This chapter is licensed under the terms of the Creative Commons Attribution 4.0 International License (http://creativecommons.org/licenses/by/4.0/), which permits use, sharing, adaptation, distribution and reproduction in any medium or format, as long as you give appropriate credit to the original author(s) and the source, provide a link to the Creative Commons license and indicate if changes were made.

The images or other third party material in this chapter are included in the chapter's Creative Commons license, unless indicated otherwise in a credit line to the material. If material is not included in the chapter's Creative Commons license and your intended use is not permitted by statutory regulation or exceeds the permitted use, you will need to obtain permission directly from the copyright holder.

Chapter 6
Wine Tourism, Circular Economy Practices and Hospitality in the Spanish Wine Industry: The Case of Bodegas Casa Sicilia Wine Restaurant

Bartolomé Marco-Lajara, Javier Martínez-Falcó, Eduardo Sánchez-García, and Luis A. Millán-Tudela

Introduction

Wine tourism (WT) is understood as a typology of tourism that can be employed as a means of regional growth, connecting primary (viticulture), secondary (wine business) and tertiary (tourism) industries.

WT is characterized by its link with nature, local gastronomy and local products, becoming one of the preferred options for travelers who wish to avoid the negative externalities associated with mass tourism. This type of tourism acts as a distribution channel for the sale of wine in the winery, generates economic wealth in the region in which it is located, and favors the care and promotion of the winemaking heritage and biodiversity, thus improving Sustainable Performance (SP), as well as the environmental management and heritage of the winery.

The WT-SP link can be enhanced by the Circular Economy Practices (CEP) developed by wineries, since they favor the optimization of resources, the reduction of raw material consumption and the use of waste, recycling it or giving it a new life to turn it into new products (Martínez-Falcó et al. 2023a). This can be translated into the improvement of SP through practices aimed at reduction, reuse and recycling, on the one hand, and in the improvement of WT activity, on the other, since CEPs act as a pole of attraction for wine tourists with a high level of environmental awareness. WT can also enhance the winery's hospitality offer, with the winery restaurant being a strategic mechanism for improving the winery's economic results and disseminating the winery's commitment to society and the environment.

B. Marco-Lajara (✉) · J. Martínez-Falcó · E. Sánchez-García · L. A. Millán-Tudela
Department of Business Organisation, Universidad de Alicante, Alicante, Spain
e-mail: bartolome.marco@ua.es; javier.falco@ua.es; eduardo.sanchez@ua.es; luisantonio.millan@ua.es

© The Author(s) 2024
M. Segarra-Oña et al. (eds.), *Managing the Transition to a Circular Economy*, SpringerBriefs in Business, https://doi.org/10.1007/978-3-031-49689-9_6

The research therefore aims to answer the following three Research Questions (RQs): (RQ1) does WT positively influence the SP of wineries? (RQ2) Do CEPs moderate the relationship between WT and SP of wineries? and (RQ3) does hospitality activity mediate the relationship between WT and SP of wineries? To answer the above RQs, a case study is conducted. This analysis has focused on the context of WT in Spain, as it represents a flourishing typology of tourism that enhances the economic and social welfare of the Iberian country (Martínez-Falcó et al., 2023b).

After this brief introduction, section "Literature Review" presents the propositions to be contrasted, section "Methodology" the steps followed to achieve the proposed research objectives, section "Results" the analysis of the interview content and, finally, section "Conclusion" the main conclusions, limitations, and future lines of research.

Literature Review

Wine Tourism and Sustainable Performance

WT can present itself as an appealing distribution channel from an economic perspective due to the fact that direct sales in wineries have become the go-to indicator for academics when measuring the economic contribution of this activity (Smyczek et al., 2020). Through this, cost savings can be achieved, as there is no need for intermediaries, and the winery is able to receive immediate financial gain as opposed to other methods. Additionally, it enhances up-selling and cross-selling opportunities and also enables wineries to build direct relationships with customers, which is a way to guarantee future sales and, at the same time, generate an emotional bond with the brand.

From the social perspective, WT could be seen as a major response to the problem of depopulation and the advancement of the welfare of society as a whole. It has been shown that WT has the potential to bring about economic growth in terms of diversification of activities which can improve the stability and conditions of work for those employed in wineries. WT can also positively influence the environment, creating externalities such as making the streets look better and providing a greater 'cultural' and 'leisure' range of activities (Duarte-Alonso et al., 2021).

From an environmental standpoint, WT can be a powerful tool to reinforce the ecological sustainability of wineries. There are both endogenous and exogenous motivating factors that can influence this merit such as the ability for a winery to set itself apart from competitors, augmenting gastronomic wealth, and protecting biodiversity. Moreover, it is likely that WT may draw in wine tourists with a sensitive view on the environment, increasing the stock of ecological knowledge within a winery. Employees in charge of such tourism activities interact with other staff to more clearly express the winery's sustainable actions, and continuously seek training to advance their knowledge of sustainable practices (Karagiannis and Metaxas, 2020).

WT is becoming an increasingly important activity to promote economic, social, and environmental development. Despite its potential, there is still insufficient research focusing on the effects of this activity on economic, social, and environmental performance in wineries. This chapter aims to contribute to this line of research by exploring the following proposition:

P1. WT Has a Positive Influence on SP of Wineries

Wine Tourism, Circular Economy, Sustainable Performance

CEPs offer an alternative to the current production and consumption model to mitigate the effects of climate change in the context of the wine industry. They consist of sharing, reusing, repairing, renewing and recycling existing materials and products, involving the three R's: recycle, reduce and reuse. Benefits to the organization associated with CEPs include increased investor interest and access to financing, higher customer satisfaction, better adaptation to environmental regulations and a strengthened reputation, amongst others (Minunno et al., 2020).

CEPs can be applied to the wine industry in many ways. For instance, green supply chain management as a form of CEP can significantly cut operating costs and reduce emissions, as well as improve the financial performance of wineries. As CEPs provide tools for environment-friendly practices such as recycling, reduction, and reuse, they can attract tourists with sustainability-conscious behaviors to wineries (Bag et al., 2022).

Investigation into the wine industry has shown that many businesses within the sector are following the principles of CEPs. Recent literature on the subject shows that PECs have the capacity to improve both WT activities and winery performance. However, to the best of the authors' knowledge, there are no previous studies that have analyzed the moderating role of PPCs in the WT-SP linkage. This presents an opportunity to further explore this relationship and contribute knowledge to related fields. Thus, this chapter proposes the following proposition:

P2. CEPs Moderate the Relationship Between WT and SP of Wineries

A Framework for Environmental Management in the Wine Tourism Sector

The framework presented in Fig. 6.1 illustrates the associations between WT, environmental management and hospitality outcomes, given that WT actions have direct implications for consumer experiences and thus implications for hospitality management.

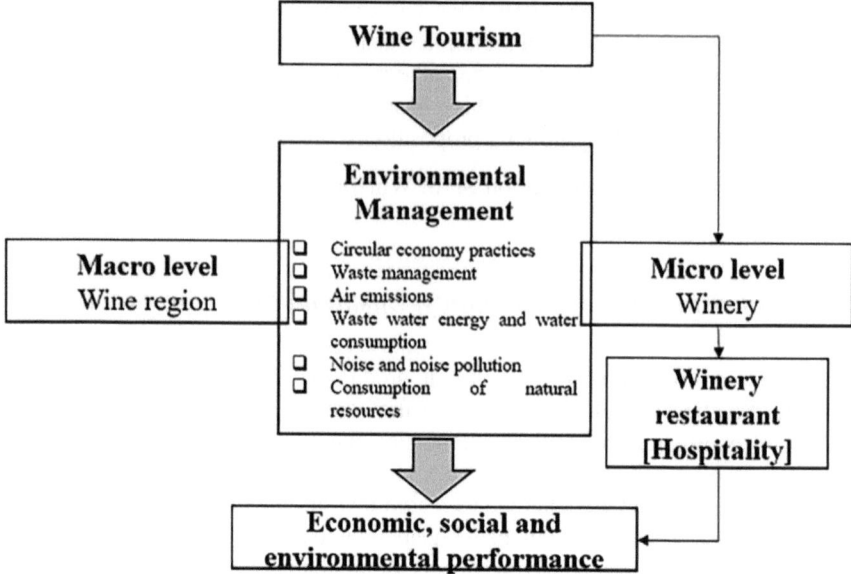

Fig. 6.1 Framework for understanding the relationship between WT, environmental management, hospitality and sustainable winery performance. Source: own elaboration

WT can catalyze wineries' environmental practices, since, on the one hand, it increases the stock of ecological knowledge held by employees by encouraging interaction between workers in order to develop WT activities satisfactorily and, on the other hand, environmental management favors the attraction of environmentally conscious wine tourists, which in turn leads wineries to intensify their efforts to protect the environment.

Such environmental management can be developed at two levels. The first level refers to the overall opportunities represented by the wine region, while the second refers to individual opportunities (the winery). On the one hand, wine regions at the aggregate level can take a number of measures to improve environmental management, including developing and promoting the adoption of sustainable agricultural practices, which helps to minimize environmental damage to the territories, such as soil erosion, water pollution, and poor air quality. These actions can be driven by the Designations of Origin in question, as they are the guarantors of the policies developed in a given wine region. On the other hand, as explained in the previous section, individual wineries can develop their environmental management, highlighting the role of CEPs, in order to improve their SP.

The framework highlights the key role of wineries in providing memorable experiences for consumers and visitors, with an implicit link to the hospitality environments in which wine is consumed. WT favors the attraction of visitors, increasing the likelihood that they will remain as diners after the activity is completed. In fact, often the WT activity itself includes, depending on the tourist package contracted by the wine tourist, lunch/dinner. This has an impact, therefore, on improving economic results by allowing the wine tasting to be combined with the winery's gastronomic offer (economic performance), generating new jobs linked to the hotel and catering industry and promoting the local gastronomy of the wine-growing areas (social performance), as well as improving the environmental awareness of diners through the explanation of the environmental management practices developed by the winery (environmental performance). Winery restaurants therefore play a crucial role in transmitting the quality of the wines, as well as the social and environmental practices developed by the wineries to the visitors who come to their premises, being these higher as wine tourism intensifies in their facilities. Based on the above ideas, we put forward the following proposition:

P3. Winery Restaurant Activities Mediate the Relationship Between WT and SP of Wineries

Methodology

A qualitative approach based on the single case study method was used for the study. Thus, although the sample was selected by non-probability sampling, it was chosen very carefully. In particular, four criteria were considered: (1) that it was a winery with WT activities, (2) that it developed and implemented actions linked to environmental management, (3) that it had a restaurant service (winery restaurant), and (4) that it was a benchmark for its commitment to society and the environment. Finally, the Casa Sicilia winery was selected for meeting the established criteria.

A triangulation using interviews, direct observation and internal and external documentation was carried out to collect data, improving the quantity and quality of the information. On July 19, 2022, a 1-hour interview was conducted with the head of WT Susana Arias Paredes, and was completed with a visit to the winery to observe the actions developed, as well as through access to internal documentation (newsletters, manual of good practices, code of ethics, etc.) and external documentation of the winery (press releases, corporate website, promotional videos, etc.).

The case study is a widely used method in social sciences, which allows us to generate new scientific knowledge. The analysis began by exploring the existing academic literature on WT, environmental management, hospitality and SP. To understand how these variables are linked, an in-depth interview was conducted at the Casa Sicilia winery and structured in four blocks linked to the variables mentioned above. This was recorded in its entirety and subsequently shared to obtain the consent of the interviewee. Finally, the interview was transcribed and content analyzed.

Results

Casa Sicilia is a winery located in the municipality of Novelda, located in the region of Medio Vinalopó. It was founded in 1707 and has its origin in the Heretat de Cesilia created by the Marquis de la Romana. At present, the estate where the winery is located is still on the outskirts of Novelda, in the same place where the heretat of the Marquis de la Romana was born, extending over four areas of the town: Alcaidías, Ledua, La Mola and Sicilia, being located in a gorge, a short distance from the hill of "La Mola", where the Sanctuary of Santa María Magdalena and the Castle of La Mola are located.

WT activities carried out by the winery make it possible to capitalize on the territory's winemaking heritage, while offering a unique experience to wine tourists who come to its facilities. Casa Sicilia promotes responsible behavior among its visitors, since it stresses the importance of responsible wine consumption during the WT activity. This awareness is of utmost importance considering that more than 90% of its visitors arrive at the winery by private car. In the words of Arias: "*During the wine tourism activity, the importance of responsible consumption and the need not to drink excessively if they have come with their private car is emphasized*".

Casa Sicilia also explains through its WT activities how it saves water and energy resources, since, on the one hand, they explain that the cultivation is rainfed, with the consequent water savings that this entails and, on the other hand, they explain how they take advantage of the open spaces (through their windows), as well as the use of low-consumption lights, thus reducing the energy consumption of the winery. In these terms Arias expresses herself in this respect: "*During the wine tourism visit, it is explained how the winery saves water resources. For example, it is explained that the crop is rainfed. Therefore, we use almost no water.*" [...] "*We try to take advantage of natural light as much as possible, creating diaphanous spaces for this purpose, as well as turning off lights when they are not necessary. Recently, we changed all the lights to low consumption.*"

WT allows the members of Casa Sicilia to interact with each other, since those in charge of this activity must be in contact with other members of the winery (winemakers, quality and environmental managers, etc.) to be able to adequately transmit to wine tourists the production process, the quality of the vintages, the certificates obtained and the awards received, among other aspects. In the same way, teamwork is encouraged in the winery and workers are trained under the pillars of sustainability, so that they can subsequently transmit the organization's awareness of environmental and territorial care. In this sense, as Arias points out: "*For the success of the wine tourism activity we have to work as a team, because we have to be in contact with all our colleagues to know what is being done in the winery and be able to transmit it properly to tourists.*" [...] "*From the winery we try to make sure that all the workers know the processes and sustainable practices that are carried out so that we can then transmit them adequately to the visitors who come to our facilities.*"

Casa Sicilia is committed to the use of indigenous grape varieties from Alicante for the production of its wines (Monastrell grapes for red wine and Moscatel for white wine), as well as organic vineyards to ensure the quality and sustainability of the wines it produces. These actions are transmitted to wine tourists so that they can learn about the winery's commitment to the use of native grapes and organic vineyards for the production of its wines. In the words of Arias: *"We use the native varieties Monastrell for the reds and Moscatel for the whites. In addition, we have two garnachas that we use for a rosé wine and a red wine that we produce."* [...] *"The entire vineyard area has been organic for several years now."* In the same way, the winery's WT manager stress the importance of preserving the vineyard landscape, given that it is a fundamental identity element to enhance the value of the wine-growing territories, also allowing to value the biodiversity that surrounds the winery, as well as the local gastronomy of the area. In this regard, it should be noted that Casa Sicilia conceives WT as a tasting activity that enhances the service offered in its restaurant, which is located on the same premises as the winery. In this way, the WT activity acts as a pole of attraction for wine tourists who can then stay to eat at the winery restaurant, being able to improve the knowledge of wine tourists about the wines offered by the winery, as well as their pairing. As Arias emphasizes: *"wine tourism allows us to have a constant flow of diners in our restaurant, allowing visitors to get to know our wines and the typical gastronomy of the region better"*.

Conclusion

This research allows confirming the research propositions raised for the case of Casa Sicilia, since it shows the role of WT in catalyzing the economic, social and environmental performance of the winery, as well as the mediating and mediating role of CEPs and winery restaurant activity, respectively, in this linkage.

WT can improve the economic performance of the winery under study by representing a distribution channel for the sale of wine, as well as improve its social and environmental performance by generating new employment in the territory where the activity is carried out, publicizing the environmental management practices carried out, disseminating the history and gastronomy of the wine-growing territory and preserving biodiversity and wine heritage.

Despite the important contributions of this chapter, it should be noted that it has limitations. In this sense, it is worth mentioning the limitation of the methodology used, given that by using the case method the results derived from the study cannot be extrapolated to the target population. In order to overcome this deficiency, as a future line of research, it is proposed to collect primary information through a questionnaire addressed to Spanish wineries in order to subsequently contrast the relationships proposed in this research by means of structural equation modeling (PLS-SEM).

References

Bag, S., Dhamija, P., Bryde, D., & Singh, R. (2022). Effect of eco-innovation on green supply chain management, circular economy capability, and performance of small and medium enterprises. *Journal of Business Research, 141*, 60–72.

Duarte-Alonso, A., Bressan, A., Kiat Kok, S., & O'Brien, S. (2021). Filling up the sustainability glass: wineries' initiatives towards sustainable wine tourism. *Tourism Recreation Research, 47*(5), 512–526.

Karagiannis, D., & Metaxas, T. (2020). Sustainable wine tourism development: case studies from the Greek region of Peloponnese. *Sustainability, 12*(12), 5223.

Martínez-Falcó, J., Marco-Lajara, B., Zaragoza-Sáez, P., & Millan-Tudela, L. A. (2023a). Do circular economy practices moderate the wine tourism–green performance relationship? A structural analysis applied to the Spanish wine industry. *British Food Journal*. https://doi.org/10.1108/BFJ-10-2022-0833.

Martínez-Falcó, J., Marco-Lajara, B., Zaragoza-Sáez, P., & Millan-Tudela, L. (2023b). Wine tourism as a catalyst for green innovation: evidence from the Spanish wine industry. *British Food Journal*. https://doi.org/10.1108/BFJ-08-2022-0690.

Minunno, R., O'Grady, T., Morrison, G., & Gruner, R. (2020). Exploring environmental benefits of reuse and recycle practices: A circular economy case study of a modular building. *Resources, Conservation and Recycling, 160*, 104855.

Smyczek, S., Festa, G., Rossi, M., & Monge, F. (2020). Economic sustainability of wine tourism services and direct sales performance-emergent profiles from Italy. *British Food Journal, 122*(5), 1519–1529.

Open Access This chapter is licensed under the terms of the Creative Commons Attribution 4.0 International License (http://creativecommons.org/licenses/by/4.0/), which permits use, sharing, adaptation, distribution and reproduction in any medium or format, as long as you give appropriate credit to the original author(s) and the source, provide a link to the Creative Commons license and indicate if changes were made.

The images or other third party material in this chapter are included in the chapter's Creative Commons license, unless indicated otherwise in a credit line to the material. If material is not included in the chapter's Creative Commons license and your intended use is not permitted by statutory regulation or exceeds the permitted use, you will need to obtain permission directly from the copyright holder.

Chapter 7
Circular Economy Practices in the Spanish Beer Industry: The Case of the Beer Producer La Somniada

Francisco Puig, Guillermo Navarro-Sanfelix, and Santiago Cantarero

Introduction

There are some processed foods that human beings have been producing and consuming for thousands of years, such as bread, wine, cheese, olive oil, or beer, and in which significant changes are currently being observed regarding their use and consumption. It is illustrated by the phenomenon of the return to the origins of the product, to *"less is more,"* to coming back to the essence of what the product itself is, and to produce more sustainably.

In this phenomenon, craft and organic beer production's strategy connects sustainability and circular economy concepts. It is based on preserving natural resources and reducing the negative environmental impacts of human activity. Such a way that the circular economy is considered an essential tool to achieve sustainability, as it reduces the pressure on natural resources and minimizes the negative environmental impacts of human activity (Pla-Julián and Guevara, 2019).

The case of beer is an example of a product with cohabitation between industrial and craft production in its transition towards a circular economy. While the business model of the former has been consolidated through the development of continuous flow production systems and economies of scale, dominating 98% of the total market, the craft is making its way towards a model based on small production batches of a beverage without additives and under a pasteurization-free process (Garavaglia and Swinnen, 2017).

This trend towards greater awareness of the environment, common to other sectors such as tourism, and with which it maintains a close relationship, has its reason in the belief that the circular economy represents a development strategy that enables economic growth while optimizing the consumption of natural resources

---

F. Puig (✉) · G. Navarro-Sanfelix · S. Cantarero
Department of Business Management, Universitat de València, Valencia, Spain
e-mail: francisco.puig@uv.es; santiago.cantarero@uv.es

© The Author(s) 2024
M. Segarra-Oña et al. (eds.), *Managing the Transition to a Circular Economy*,
SpringerBriefs in Business, https://doi.org/10.1007/978-3-031-49689-9_7

Fig. 7.1 Visual depiction showcasing various styles of artisanal and environmentally-friendly beer production

through a profound transformation of production chains, consumption patterns and the redesign of industrial systems (Segarra-Oña et al., 2021). The relevance of designing for eco-effectiveness, i.e., seeking products with a positive impact instead of focusing on reducing negative impacts, can generate business opportunities with a positive social and environmental impact (increased employment and economic competitiveness, resource consumption and waste prevention) (Pla-Julián and Guevara, 2019). However, in countries such as Spain, the production of ecological beer is anecdotal, with exceptions such as the brewery La Somniada and its different types of beers L'Audaç (Fig. 7.1).

Given the benefits that sustainable production has for society, as well as the challenges that this model entails for its companies, in this chapter, we will study the model implemented by this company and its viability. To this end, the following chapter contextualizes the brewing industry in Spain and then describes the origins, business model, and circularity practices of the brewing cooperative La Somniada. It concludes with a discussion of the main challenges this firm will face to consolidate its commitment to sustainability. The conclusions of this analysis may help entrepreneurs and agents of the economy to develop actions that can be more effective.

7 Circular Economy Practices in the Spanish Beer Industry: The Case of... 71

Table 7.1 Data relative to the size of the companies

	2017	2018	2019	2020	2021
Turnover (th. €)	229,4	247,3	274,2	236,0	261,6
Number of firms	150	180	205	237	253
Employees	480	501	600	750	761
Average employees	3,2	2,8	2,9	3,2	3,0

Source: Own elaboration base on SABI (2023)

Industrial and Craft Beer in Spain

In general, the Spanish beer industry is in good health (ISCE, 2021). Over the last ten years, annual per capita consumption has remained around 50 liters. Although it is far from the 129 liters consumed by the Czechs or the 92 liters consumed by the Poles, it is somewhat higher than in neighboring countries such as Portugal, France, or Italy (STADISTA, 2023).

The Spanish beer consumer is also characterized by being adventurous, i.e., open to trying new beers and showing a variety of preferences in terms of beer styles. As a result, Spanish beer culture has been diversifying, and there is a greater willingness to experiment with other beers, such as craft beers (Alimarket Report, 2023).

Regarding the consumption of craft beer, although there are no specific data, the Spanish Brewers Association (ITCA, 2020) estimates that the production and consumption of craft beer in Spain has an annual growth rate of close to 20% (according to our estimation, from 150 in 2017 to 253 in 2021). This fact is evidenced by the continuous creation of new craft breweries and the evolution of their average income. According to data extracted in March 2023 from the SABI[1] database, other characteristics of this segment are the small average size of the companies, with a turnover of around 260,000 € and a workforce of around three employees (Table 7.1).

Approximately 98% of the beer consumed in Spain is produced industrially, and the remaining 2% is artisanal. The 65% of the industrial beer production is concentrated in four manufacturers (Grupo Mahou-San Miguel, Grupo Damm, Grupo Heineken España, and Corporación Hijos de Rivera), while craft beer is atomized and dispersed throughout Spain (ITCA, 2020).

Regarding the location of the craft beer firms, it is in Catalonia and Andalusia, where the most significant number of them are located (26% and 12%, respectively). The Valencian Community is home to approximately 8% (Fig. 7.2). This geographical dispersion indicates that, in Spain, as in other European countries, this industry focuses on producing beer for local consumption.

---

[1] The SABI database refers to the "System for Analysis of Iberian Balances". It is a financial database that contains comprehensive information on companies in Spain and Portugal. It supplies detailed financial data, including balance sheets and income statements. It is provided by Bureau van Dijk.

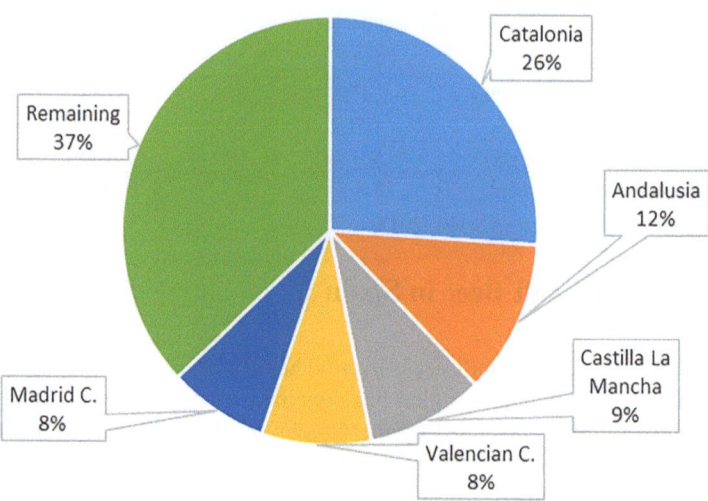

Fig. 7.2 Visual depiction showcasing two of its best well-known varieties of beers (Primavera and Serrana). Source: Own elaboration base on SABI (2023)

However, beyond its demographics, the most distinguishing feature of craft beer is its innovative and creative spirit. Brewers constantly experiment with new ingredients, brewing techniques, and beer styles to offer new flavor proposals. While industrial beers prioritize homogenization, craft beer is about innovation and creativity. This is due to the presence of live yeast and because it is produced through a natural process from the grain, without using extracts or products different from the traditional ones (water, yeast, hops, and the cereal to make the malt) (Garavaglia and Swinnen, 2017).

Consequently, it is common for new proposals, such as craft and organic beers, to appear. In this way, artisans add to their commitment to flavor, sustainable production, circularity, and the constant development of new recipes as naturally as possible. In this case, producers show a trend towards waste reduction in production plants, optimization in water use, intensive use of renewable energies, and better use of packaging (reducing plastic and using recycled cardboard).

However, despite the potential of the circular economy in this sector, it is very little developed in Spain. In line with Ormazabal et al. (2018), the main barriers that could discourage the implementation of the circular economy in these businesses would be the following:

- High initial costs due to the assets' specificity and limited financial resources.
- Infrastructure constraints: Implementing circular practices often depends on the availability of adequate infrastructure, such as recycling facilities or by-product collection points.
- Regulatory barriers: Some regulations or standards may hinder the implementation of circular practices.

- Availability of suppliers and partners: Craft beer producers need to collaborate with suppliers and other value chain actors who share the circular economy vision to implement circular practices.
- Awareness and education: The circular economy is a relatively new concept, and there may be a lack of widespread awareness and understanding of its benefits and applications.

The Brewing Cooperative La Somniada

Origins and Context

The company's founders are Gabrielle, David, and Javier, who met at a university forum and shared an entrepreneurial spirit, respect for nature, and concern for recovering and creating traditional products. Gabrielle is the master brewer and the partner most involved in craft beer production, while David works on the project's expansion and commercial side. Finally, Javier is present in the brewery as support in production, bureaucratic management, and design.

In 2021, they decided to give the green light to the project and chose Soneja, a small town in the Alto Palancia region in the Castellón (Spain) province. The reasons for this decision lie in its proximity to other local producers and markets and the potential for collaborative relationships with other small businesses and farmers in the area. Its legal form is the cooperative, and it is justified because it is the one that most closely aligns with its values of sustainability, the creation of craft beer, and respect for the rural world (Cantarero et al., 2013).

Cooperativa La Somniada focuses on satisfying beer-educated consumers looking for a unique experience. These consumers are willing to pay higher prices for craft beers, seeking specific characteristics that add value to their experience (Fig. 7.3 shows two of its varieties). They would prefer to try something other than the traditional beer taste and prefer to try different beers to broaden their knowledge.

The Business Model

La Somniada's value proposition is based on clearly identifying the target market, adding value to its beers through local seasonal ingredients and spices ecologically and sustainably, and being clear about its essential resources and relationships with other agents. Other characteristics of its business model are summarized in Table 7.2.

This is carried out through the development of three main actions:

Fig. 7.3 Visual depiction showcasing two of its best well-known varieties of beers (Primavera and Serrana)

Table 7.2 Summary of La Somniada Cooperative's business model

Who is it aimed at?	Customers educated in beer culture. Close to the place of production.
The origin of the of beer	The value of territory as a brand. Synergies with local producers and artisans. Maximum respect for the environment.
Sustainable distribution and sourcing	Proximity both in sourcing and in distribution. Higher economic cost of raw materials. Lower cost of uncertainty and quality.
Sources of income	Restaurants and specialized stores. Customers of your website. Visitors to the brewery. Bulk sales.
Dealing with customers	Stable customers (Restaurants, stores, and bulk sales). Occasional customers (Online store and visitors to the cooperative).
Key resources	The particularity of its human capital. The intensity of its relational capital. The intensity of its values with the environment. The legal form of the enterprise.
The role of the institutions	Artisan companies in the environment. The support of public institutions.

Source: Own elaboration based on interviews

1. Close collaboration with local producers to take advantage of the waste generated in beer production, which is converted into raw material for nearby livestock farmers or sold in the cooperative.
2. Sourcing raw materials from nearby producers to reduce transportation costs and environmental impact.
3. Commitment to proximity customers who share the same philosophy and whose distribution channels serve to offer their products and create fully comprehensive artisanal experiences for end consumers.

Circular Economy Practices

Circularity practices based on reactive (e.g., in response to government regulations) or proactive (e.g., business opportunity) strategies can be observed in business reality. In many cases, companies may combine both reactive and proactive reasons. In the latter case, firms recognize the potential benefits of the circular economy and actively pursue innovation and competitive advantage by adopting circular approaches in their operations, product design, and supplier and customer relationships (Ormazabal et al., 2018). The transition process is not sequential, allowing one to identify realized practices and others with the desire for realization.

In the case analyzed, Somniada is characterized by a commitment to sustainable distribution and sourcing in the group of realized practices. La Somniada takes sustainability into account when designing its distribution and procurement channels. Due to the high cost of moving water-based products such as beer and the associated fuel emissions, the brewery sells mainly to nearby customers in the province of Castellón. Similarly, the company purchases its raw materials from nearby suppliers to avoid transportation-related costs and uncertainty in the quality of the purchased product.

To carry out these practices, it relies on specialized human capital and adopts a holistic approach that considers the interrelationships within and between sectors, value chains, and ecosystems. This approach allows it to obtain income from two sources: (a) direct or related to restaurants and specialty stores in the area, direct sales on its website, visitors to its facilities, and bulk sales as a private label, and (b) indirect and linked to the sale of surplus raw materials to local livestock and bakeries, as well as receiving funds from rural development programs.

In addition, the cooperative's worker-members are also highly trained in beer production and biology, enabling them to develop artisanal processes and experiment with new beer formulas while using environmentally friendly machinery. The firm maintains a solid commitment to sustainability and respect for the rural world, reflected in its relations with organizations and associations of neighbors and businesses in the surrounding area.

Those that have yet to be carried out and would like to be carried out would be those related to improving processes and efficiency. In this respect, implementing a bottle washing and recycling system to ensure maximum reuse of raw materials seems logical. However, this activity must be carried out considering a cost structure that allows them to be viable and involves a more active search for economies of scale. To this end, the company should consider certain external aspects, such as the demographics of its environment, potential expansion to other provinces, and public initiatives in the form of quality seals.

Conclusions

Throughout the chapter, we have seen that sustainable development policies should integrate circular thinking to address the United Nations SDGs set for the 2030 Agenda. In the case of products such as beer, which is strongly linked to the tourism sector, this challenge should be achieved by reinforcing the social aspects of the production process. In other words, it raises awareness among consumers and other agents in the value chain. In our analysis, we have seen that achieving this requires an understanding of complex intersections and organizational and technological obstacles that require the active participation of formal and informal institutions at different levels and entrepreneurial resources and capabilities.

In a competitive context characterized by a vast majority consuming industrial beer (98%), high concentration (four producers produce more than 65% of the industrial beer), average per capita beer consumption, low beer culture, and low entry barriers to the sector, there is a growing number of craft producers and a substantial increase in the sector's rivalry. This diversification and scenario of the sector towards craft beer makes it possible to contemplate the potential of organic beer within its offer.

Faced with this new socioeconomic and competitive reality, the cooperative La Somniada's differentiating strategy should be analyzed. As a result of the various interviews, complemented by the study of secondary sources, we have observed that its focus on craft and organic production seems wise. Moreover, the business model on which it has based its development aligns with this decision.

However, the project faces significant challenges due to its small size and its disadvantages in addressing sustainability plans. For example, large companies focus on implementing technological innovations directly influencing the profit and loss account, being little related to the production process (energy efficiency, waste reduction, recycling, or sustainable packaging). In contrast, craft breweries focus on efficiently managing the production process with the environment (nearby raw materials, local distribution, water conservation), which makes them lose price competitiveness and affects their profitability.

References

Alimarket Report. (2023). *Informe 2022 del sector de cervezas en España*. https://www.alimarket.es/alimentacion/informe/353903/informe-2022-del-sector-de-cervezas-en-espana

Cantarero, S., González-Loureiro, M., & Puig, F. (2013). El efecto "economía social" en la supervivencia empresarial. *CIRIEC-España, Revista de Economía Pública, Social y Cooperativa, 78*, 175–200.

Garavaglia, C., & Swinnen, J. (Eds.). (2017). *Economic perspectives on craft beer: A revolution in the global beer industry*. Springer.

ISCE (Informe socio económico del sector de la cerveza en España). (2021). *Ministerio de Agricultura, Pesca y Alimentación*. Avaliable from https://cerveceros.org/uploads/62cfc9469b35d__InformeSocioeconomico_Cerveza2021.pdf

ITCA (Informe técnico de la cerveza artesana e independiente española). (2020). Available from https://aecai.es/wp-content/uploads/2021/12/informe-cerveza-artesana-espana.pdf.

Ormazabal, M., Prieto-Sandoval, V., Puga-Leal, R., & Jaca, C. (2018). Circular economy in Spanish SMEs: Challenges and opportunities. *Journal of Cleaner Production, 185*, 157–167.

Pla-Julián, I., & Guevara, S. (2019). Is circular economy the key to transitioning towards sustainable development? Challenges from the perspective of care ethics. *Futures, 105*, 67–77.

Segarra-Oña, M., Peiró Signes, A., García Meseguer, C. A., & Sánchez-Planelles, J. (2021). Análisis de la eco-innovación, circularidad y acciones simbióticas. Una reflexión aplicada al sector turístico valenciano. *Economía Industrial, 418*, 89–96.

STADISTA. (2023). *Volumen de consumo per cápita de cerveza en Europa en 2021, por país*. https://es.statista.com/estadisticas/1147758/consumo-per-capita-.

Open Access This chapter is licensed under the terms of the Creative Commons Attribution 4.0 International License (http://creativecommons.org/licenses/by/4.0/), which permits use, sharing, adaptation, distribution and reproduction in any medium or format, as long as you give appropriate credit to the original author(s) and the source, provide a link to the Creative Commons license and indicate if changes were made.

The images or other third party material in this chapter are included in the chapter's Creative Commons license, unless indicated otherwise in a credit line to the material. If material is not included in the chapter's Creative Commons license and your intended use is not permitted by statutory regulation or exceeds the permitted use, you will need to obtain permission directly from the copyright holder.

Chapter 8
Good Practices of Circular Economy in Tourism in Castellón

Andrei Serbanescu, Luís Martínez Cháfer, and Teresa Martínez Fernández

Introduction

Tourism is one of the fundamental pillars of the Spanish economy and its importance is undoubted. Spain is one of the most popular tourist destinations in the world, with tourism reaching an estimated figure of 159,490 million euros in 2022 (Exceltur, 2023), which represents 12.2% of the country's GDP. These figures are much closer to the reality prior to the global pandemic, considering that the figure of 157,355 million euros was reached in 2019, which represents 12.6% of the country's GDP, while in 2021 97,126 million euros were collected (INE, 2023).

Thus, this activity represents a key sector for the country, both from the economic and social point of view, since it generates a large amount of employment and wealth. In addition, tourism contributes to promoting economic development in regions with less activity and to diversifying the country's economy.

But tourism's impact is not exclusively related to positive effects in the economic and social sphere. Tourism has a direct impact on the environment and is a great challenge in terms of waste and pollution management given the high number of visitors that Spain receives each year (Sgambati et al., 2021; Renfors, 2022). It is therefore essential to promote sustainable and responsible tourism.

Given that the circular economy (CE) is an economic model that seeks to minimize waste and maximize the reuse of resources, it can be fundamental for the tourism sector. The World Travel & Tourism Council's (WTTC) study on circular economy highlights that CE can help tourism businesses reduce costs and improve their long-term sustainability, while reducing the environmental footprint (World Travel & Tourism Council & Harvard T.H. Chan School of Public Health, 2022).

A. Serbanescu (✉) · L. M. Cháfer · T. M. Fernández
Department of Business Administration and Marketing, Universitat Jaume I, Castellón, Spain
e-mail: al385173@uji.es; chafer@uji.es; tmartine@uji.es

© The Author(s) 2024
M. Segarra-Oña et al. (eds.), *Managing the Transition to a Circular Economy*,
SpringerBriefs in Business, https://doi.org/10.1007/978-3-031-49689-9_8

The concept of CE can be understood as an economic system that replaces the 'end-of-life' concept with reducing, alternatively reusing, recycling, and recovering materials in production/distribution and consumption processes (Ellen MacArthur Foundation, 2013). It operates at the micro level (products, companies, consumers), meso level (eco-industrial parks) and macro level (city, region, nation and beyond), with the aim to accomplish sustainable development, thus simultaneously creating environmental quality, economic prosperity and social equity, to the benefit of current and future generations (Kirchherr et al., 2017; Prieto-Sandoval et al., 2018).

Therefore, CE follows a cradle-to-cradle perspective, where materials are viewed as nutrients circulating in healthy and safe metabolisms. This design enables the creation of wholly beneficial industrial systems driven by the synergistic pursuit of positive economic, environmental, and social goals (Braungart et al., 2007). Also, a review on CE made by Ghisellini et al. (2016) states that CE allows a reduction of the costs for the companies and municipalities due to a reduction of the problem of waste management, as well as to a reduction of the externalities for the society (lower pollution), new job opportunities and increased welfare for low-income households. Regarding employment creation, the implementation of a CE could generate up to 700,000 additional jobs in the tourism sector by 2040, according to a report by the Ellen MacArthur Foundation (2021).

But, on the other hand, to achieve a CE, it is necessary for both producers and consumers to take an active role in recycling or reusing products, moving away from the passive "throwaway" mentality that characterizes the linear economy (Shah, 2014; Rodríguez et al., 2020). Furthermore, given that economic expansion cannot continue indefinitely, it is crucial to view the CE as a shift toward a novel business model that decouples wellbeing from resource consumption. By promoting less resource utilization and greater wellbeing, CE could facilitate the transition to a degrowth trajectory, which appears increasingly necessary in developed economies that have already exceeded ecological thresholds (Ghisellini et al. 2016; Kerschner 2010).

In conclusion, as opposed to the linear "one use only" model, the CE focuses on creating systems in which products, materials and resources maintain their value and usefulness as long as possible, avoiding the generation of waste and minimizing the extraction of raw materials. Therefore, its importance in today's society and tourism is very high, not only for being environmentally positive, but also for generating important economic and social benefits.

Thus, CE represents a real and necessary alternative to build a fairer, more sustainable, and environmentally friendly tourism model. That is why it is essential to promote the transition towards a CE both from the private and public spheres and raising awareness about its importance.

All this leads to the need to observe which specific actions are being carried out in this field, in order to learn about companies and citizens' current position regarding CE in tourism.

In this chapter, we will make a compilation and analysis of good practices of CE performed in the tourism sector of the province of Castellón. To accomplish this, we will mainly rely on a study carried out by the Polytechnic University of Valencia, in collaboration with the Universitat Jaume I of Castellón and the University of Alicante within the InnoEcoTur project.[1] We will also outline a specific example of a hotel in Castellón (Spain).

Castellón's Tourism Industry[2]

The province of Castellón, located on the east coast of the Valencian Community in Spain, is a popular tourist destination and of great importance for the region.

According to the report of the Evolution of Tourism Activity in the Province of Castellón (GVA, 2021), 3.3 million tourists visited the province in 2021, a 35.3% higher amount than the previous year. Regarding the figures of hotel establishments in the province, these registered significant year-on-year growth in 2021, both in the number of travelers (78.2%), and in the number of overnight stays (94.4%).

However, the comparison of 2021 and 2020 should be carefully analyzed, since both years were importantly affected by the global pandemic, compared to usual figures (for example, 4.4 million tourists arrived in the province in 2019). In any case, this also indicates that the sector has been able to recover after the situation caused by the pandemic.

The province has a wide variety of tourist attractions, from beaches and natural landscapes to historic cities and architectural monuments.

The Costa Azahar is one of the main tourist attractions in the region, since it offers a wide variety of options for tourists, from urban beaches to virgin coves surrounded by nature. In addition, the coast has many tourist services such as hotels, restaurants, bars and stores, making it an attractive option for those looking for a complete holiday.

Other tourist attractions in the province of Castellón are its charming villages and towns, such as Peñíscola, Morella or Vilafamés. Also, it has a wide range of rural and active tourism since its interior territory is full of natural parks and hiking trails.

It is also an ideal destination for gastronomic tourism lovers, since the province has a rich variety of typical dishes and local products, such as olive oil, citrus fruits, wines, or cheeses.

Thus, tourism is a key sector in the economy of Castellón throughout the year, not only during summer months. In addition, promoting sustainable tourism is a key

---

[1] Creation of an Innovation Platform for Promoting and Implementing a Circular Economy Strategy in the Tourism Sector of the Valencian Community (InnoEcoTur). https://innoecotur.webs.upv.es/

[2] This section has been written using information available on the following websites, accessed in May 2023:
  https://www.castellonturismo.com/
  https://turismodecastellon.com/es/
  https://www.castellonvirtual.es/
  https://www.valenciabonita.es/2021/07/09/descubre-interior-castellon/

objective for the future development of the province, favoring natural and cultural resources' protection while respecting the environment and local population.

Focus Groups[3]

The purpose of the InnoEcoTur project's report (InnoEcoTur, 2022) was to gather information on the requirements of companies in the tourism sector to adopt the values and principles of Circular Economy (CE). This was achieved through the organization of three focus groups, each conducted in a different province of the Valencian Community, over the period from October 2021 to January 2022. By employing this method of information gathering, the aim was to ascertain the participants' viewpoints regarding the implementation of CE within their respective companies, as outlined by Krueger and Casey (2015).

For the realization of these sessions, a profile of participants was concretized composed of high-level executives in their companies, with the capacity to make decisions on the transition to the CE; of hotel companies, restaurants, and suppliers of both throughout the Valencian Community. In addition, companies of considerable size have been selected, to ensure that they have experience in the transition to the circular economy. Specifically, the companies[4] based in Castellón that participated are Hotel Voramar, Hotel del Golf, Idear Ideas, Grupo DeCasa & Congelados DIL, Marabrasa and Pou de Beca.

Given that the focus groups were conducted within the Valencian Community, it is important to note that in this chapter, we will specifically highlight the notable circular economy practices identified in the province of Castellón. Nevertheless, obtaining information at a broader level does not imply that we cannot recognize that most of the conclusions and practices mentioned in the other two provinces can be potentially applied to the situation in Castellón as well.

In addition to this study, in order to try to reflect reality as close as possible to specific CE practices' reality, we will also collect information from other complementary sources of information such as: specialized publications, touristic companies' websites, local press reports, scientific publications, etc.

---

[3] This part of the chapter was based on the INNOECOTUR report, where the focus groups process is fully explained. It is accessible through: https://innoecotur.webs.upv.es/primer-informe-de-necesidades-del-sector-turistico-para-la-transicion-a-la-economia-circular/

[4] We would like to thank all the participants who took part in this dynamic, for sharing with us their knowledge and experiences in the transition towards the circular economy in the sector in the province of Castellón.

Circular Economy Practices in Castellón's Tourism Industry

The transition towards a CE requires a change in society's mindset and consumption habits at large. Given their complexity, the demands towards CE may seem generic or unclear, which makes their practical implementation difficult. This is why it is essential to translate these demands into concrete and tangible concepts so that they can be understood and accepted by society and to achieve greater public awareness and commitment to the CE. In the field of tourism companies, this becomes even more important, as they are dynamic agents reflecting where both direct and indirect actions that facilitate the transition towards a CE should be directed.

Thus, some examples that we have been able to identify of good circular economy practices carried out by companies in Castellón are:

- Employee training and awareness-raising on energy saving and circular economy.
- The reconversion of worn-out hotel sheets into uniforms for their staff.
- Renewal of electrical appliances, such as air conditioners, to replace them with new, more energy-efficient models.
- Measurement of the carbon footprint and the consequent establishment of actions to reduce it.
- Use of water and light sensors in hotels and restaurants.
- Creation of own reusable packaging to be used in relations between hotels and restaurants and their suppliers.
- Purification, use and reuse of water from showers and toilets in hotels.

During the focus group discussions on CE in the tourism industry of Castellón, it became evident that some participants expressed significant concerns regarding the challenges of implementing CE measures. A prominent concern revolved around the financial barriers associated with adopting sustainable practices. Participants highlighted the higher upfront costs and ongoing expenses involved in transitioning to circular business models. These cost barriers often deterred businesses from taking action, particularly smaller establishments with limited resources. Additionally, participants expressed a sense of frustration regarding the lack of consumer appreciation and understanding of CE practices. Despite their efforts to adopt sustainable measures, businesses reported a reduced competitiveness in the market due to a perceived lack of demand for eco-friendly products and services. This consumer indifference hindered the broader adoption of circular economy practices within the tourism industry, making it difficult for businesses to justify investing in sustainable initiatives.

However, there is one establishment whose involvement has stood out above and beyond the efforts and actions we have compiled. This is the Hotel Voramar in Benicàssim, whose participation in this area is based on a strategy centered on the Economy for the Common Good (ECG), a similar and complementary concept to the CE.

This case study is not intended to be an exhaustive analysis, but a concrete positive example to show other actors in the tourism industry the way to sustainability through circularity (Voramar, 2015; Ruiz-Carmona, 2018).

Following the SEGITTUR (2022) report, observed good practices are summarized in a simple fact sheet along with the identification of its circular business model, based on the proposal of Arponen et al. (2018).

Hotel Voramar

Entity	Hotel Voramar
Location	Benicassim, Castellón, Spain
Activity	Accommodation and catering
Business model	Circular sourcing, green procurement, inter-agent collaboration, reverse logistics
Rs	Redesign/Reduce/Reuse/Repair/Recycle
Description	

It is a company that aims to be excellent, with a high level of awareness of the environment, people and its surroundings. Despite obtaining a score of 40.7% in its 2015 Common Good Balance Sheet, it has obtained very good results in indicators such as fair distribution of the work volume, fair distribution of income, and social transparency and participation in decision-making
Its mission is defined as "to contribute to the wellbeing of all people, providing them with accommodation and catering services of the highest quality, efficiency and sustainability"
"Our interest is that Voramar is a pleasant, warm place, and in general, a refuge for all the people who wish to visit us. To provide sensations; to inspire moments of joy and happiness among friends, families and co-workers and to create values; values of respect and commitment that can serve as an example and help to create healthy and responsible companies"
On the other hand, its vision is "to repay our customers, our team, our suppliers, shareholders and society in general for their trust and their time".
Thus, we can highlight different CE actions carried out by this hotel, such as:
Establishment of criteria for the selection of sustainable suppliers and products, including proximity of the product, environmental certifications of the supplier, returnability of packaging used, etc.
Introduction of ethical marketing techniques, mainly through honesty in communications and the promotion of positive stereotypes, protecting the interests of individuals and disadvantaged groups
Supporting events and actions aimed at promoting local culture, environmental protection values and social commitment. To this end, they actively collaborate in numerous events in the municipality by means of a financial contribution or by providing free accommodation or catering services
Energy rehabilitation of buildings, mainly through the exchange of cold/heat with groundwater
Installation of solar and photovoltaic panels
Use of electric vehicles

WEB	https://www.voramar.net/

Therefore, we can affirm that the Voramar hotel follows the three principles on which circularity is based, disseminated by the MacArthur Foundation[5] to support the transition to CE. They focus on (1) the elimination of waste and pollution, (2) the circularity of its products and materials, and (3) the regeneration of nature. Moreover, they focus on five of the seven "Rs" of CE from a business point of view: Redesign, Reduce, Reuse, Repair and Recycle.

Conclusions

With this chapter we have tried to show a set of actions in the Circular Economy so that other agents in the tourism industry can benefit from these experiences, and thus join the desired transition to the CE and, consequently, towards greater sustainability. We understand that a collection and exchange of information through good practices in CE among the different actors involved is a fundamental tool to foster the transition from the current model to the CE model (SEGITTUR, 2022).

This chapter has provided a comprehensive overview of several commendable examples of circular economy practices implemented by companies in Castellón. These examples include employee training and awareness-raising on energy conservation and circular economy principles, repurposing worn-out hotel sheets into staff uniforms, upgrading to energy-efficient electrical appliances, measuring and reducing carbon footprints, utilizing water and light sensors, creating reusable packaging for hotel-restaurant-supplier relationships, and implementing water purification and reuse systems in hotels. Notably, the Hotel Voramar in Benicàssim has emerged as a standout establishment, going above and beyond in their commitment to the CE. By adopting a strategy rooted in the Economy for the Common Good (ECG), they have exemplified an approach that aligns with and complements the principles of circularity. These examples serve as inspiring models of sustainable business practices and highlight the potential for widespread adoption of CE principles in Castellón and beyond.

However, in this chapter we also acknowledge the existence of certain barriers to the implementation of CE practices, including: the high costs associated with carrying out these practices, lack of training, knowledge, and awareness about CE (both among customers and companies themselves), institutional and administrative barriers, and lack of incentives.

Regarding this, it is worth mentioning that a somewhat pessimistic perspective is evident among entrepreneurs in the sector. It has even been mentioned that these barriers create a disadvantaged situation compared to other competitors in the sector, highlighting the significant impact these practices have on companies' profit margins. Additionally, there seems to be a lack of interest from consumers towards CE practices, which further hampers their implementation. Addressing these concerns

---

[5] https://ellenmacarthurfoundation.org/

and raising awareness among both businesses and consumers will be crucial to overcoming the challenges and fostering a more circular and sustainable tourism sector in Castellón.

In this vein, we also need to highlight a topic that was subtly mentioned in the previous section regarding specific practices carried out in the province. This is, the difficulty of accessing information regarding these specific practices, either due to a lack of publicity or because there is indeed a limited number of practices. It should be noted that despite conducting an exhaustive search for information in this area, we were unable to compile an extensive list of such practices.

Furthermore, during this information search process, we have observed that while it is challenging to find information about specific practices implemented by companies, it is possible to find relatively abundant information and publications about claims (rather than facts) of a strong awareness and commitment from many companies in the sector regarding CE, sustainability, or carbon footprint. Numerous publications highlight certifications, awards, and recognitions bestowed upon these companies. As a result, all of this leads us to believe that there is a significant lack of specificity in terms of actual practices, even though many companies easily assert that they are doing great work in this field.

Thus, the difficulty in observing and assessing the practices effectively implemented in this area presents a certain contradiction. It is worth considering which factor has the greatest impact on the fact that it is so challenging to compile a list of CE practices, whether it is a problem of publicity or rather a lack of actual implementation of these practices in the sector.

Therefore, it is important to conduct future analysis to understand the actual efforts in the field of circular economy. In this regard, it might be necessary for tourism companies in the province of Castellón to clearly communicate the practices they are carrying out, and to promote greater transparency and societal involvement in evaluating these practices.

Acknowledgement This work was supported by MCIN / AEI /10.13039/501100011033 / FEDER, UE under Grant PID2021-126516NB-I00.

References

Arponen, J., Juvonen, L., & Vanne, P. (2018). Circular economy business models for the manufacturing industry. *Circular economy playbook for finnish SMEs*. SITRA.

Braungart, M., McDonough, W., & Bollinger, A. (2007). Cradle-to-cradle design: Creating healthy emissions – a strategy for eco-effective product and system design. *Journal of Cleaner Production, 15*(13–14), 1337–1348., ISSN 0959-6526. https://doi.org/10.1016/j.jclepro.2006.08.003

Ellen MacArthur Foundation. (2013). *Towards the circular economy*. Ellen MacArthur Foundation. https://www.ellenmacarthurfoundation.org

Ellen MacArthur Fundación. (2021). *Objetivos universales para políticas de economía circular*. Habilitando una transición a gran escala. https://archive.ellenmacarthurfoundation.org/assets/downloads/ES-Objetivos-universales-depoli%CC%81ticas-para-la-economi%CC%81a-circular.pdf

Exceltur / N°83 Enero 2023/ Valoración turística empresarial de 2022. https://www.exceltur.org/wp-content/uploads/2023/01/Informe-Perspectivas-N83-Balance-del-ano-2022-y-expectativas-para-2023.pdf Consultado en mayo 2023.

Ghisellini, P., Cialani, C., & Ulgiati, S. (2016). A review on circular economy: The expected transition to a balanced interplay of environmental and economic systems. *Journal of Cleaner Production, 114*, 11–32., ISSN 0959-6526. https://doi.org/10.1016/j.jclepro.2015.09.007

GVA. (2021). Informe sobre la evolución de la actividad turística de la provincia de Castellón. Estadístiques de Turisme de la Comunitat Valenciana. https://www.turisme.gva.es/tcv/tcv2021/3_Castellon_2021c.pdf

INE: Cuenta satélite del turismo de España. https://www.ine.es/dyngs/INEbase/es/operacion.htm?c=estadistica_C&cid=1254736169169&menu=ultiDatos&idp=1254735576863 Consultado en mayo 2023.

InnoEcoTur. (2022). *Primer Informe de necesidades del sector turístico para la transición a la economía circular*. https://innoecotur.webs.upv.es/primer-informe-de-necesidades-del-sector-turistico-para-la-transicion-a-la-economia-circular/

Kershner, C. (2010). Economic degrowth vs. steady-state economy. *Journal of Cleaner Production, 18*, 544–551.

Kirchherr, J., Reike, D., & Hekkert, M. (2017). Conceptualizing the circular economy: An analysis of 114 definitions. *Resources, Conservation and Recycling, 127*, 229.

Krueger, R. A., & Casey, M. A. (2015). *Focus groups. A practical guide for applied research*. Sage.

Prieto-Sandoval, V., Jaca, C., & Ormazabal, M. (2018). Towards a consensus on the circular economy. *Journal of Cleaner Production, 179*, 605–615. https://doi.org/10.1016/j.jclepro.2017.12.224

Renfors, S. M. (2022). Circular economy in tourism: Overview of recent developments in research. *Matkailututkimus, 18*(1), 47–63.

Rodríguez, C., Florido, C., & Jacob, M. (2020). Circular economy contributions to the tourism sector: A critical literature review. *Sustainability, 12*, 1–27. https://doi.org/10.3390/su12114338

Ruíz Carmona, O. (2018). In peace with capital? An alternative to the current capitalist system. *Treball Final de Grau en Administració d'Empreses*. Universitat Jaume I. http://hdl.handle.net/10234/176885

SEGITTUR. (2022). *Guía práctica para la aplicación de la economía circular en el sector turístico en España*. https://www.segittur.es/sala-de-prensa/informes/guia-practica-para-la-aplicacion-de-la-economia-circular-en-el-sector-turistico-en-espana/

Sgambati, M., Acampora, A., Martucci, O., & Lucchetti, M. C. (2021). *The integration of circular economy in the tourism industry: A framework for the implementation of circular hotels* (Vol. 5, p. 95). University of South Florida (USF) M3 Publishing.

Shah, V. (2014). The circular economy's trillion-dollar opportunity. Retrieved from https://www.ecobusiness.com/news/circular-economys-trillion-dollar-opportunity/. Accessed Apr 2023.

Voramar. (2015). *Balance del Bien Común*. https://www.voramar.net/app_voramar/dossier/dossierEBC2015.pdf

World Travel & Tourism Council, & Harvard T.H. CHAN. (2022). *Economy circular*. https://wttc.org/Portals/0/Documents/WTTC-Harvard-LearningInsight-CircularEconomy.pdf

Open Access This chapter is licensed under the terms of the Creative Commons Attribution 4.0 International License (http://creativecommons.org/licenses/by/4.0/), which permits use, sharing, adaptation, distribution and reproduction in any medium or format, as long as you give appropriate credit to the original author(s) and the source, provide a link to the Creative Commons license and indicate if changes were made.

The images or other third party material in this chapter are included in the chapter's Creative Commons license, unless indicated otherwise in a credit line to the material. If material is not included in the chapter's Creative Commons license and your intended use is not permitted by statutory regulation or exceeds the permitted use, you will need to obtain permission directly from the copyright holder.

Part III
Research, Innovation, Competitiveness and Production

Chapter 9
Importance of Culture and Innovation in Behaviors Towards the Circular Economy in Spanish Hotels

Bartolomé Marco-Lajara, Mercedes Úbeda-García, Esther Poveda-Pareja, and Encarnación Manresa-Marhuenda

Introduction

Tourism activity has been the subject of debate in recent times due to its economic importance and the sustainability problems that this economic activity entails: CO_2 emissions, excessive consumption and waste of water, energy, food and other resources due to practices and technologies unsustainable in the hospitality sectors (Bruns-Smith et al., 2015; Gössling et al., 2013; Hall, 2019; Koçak et al., 2020).

The debate is centered in the current underlying model of planning and growth (Hall, 2019; Higgins-Desbiolles et al., 2019; Sharpley, 2020) that has led to environmental degradation. Therefore, many companies face the inevitable pressure to adopt their business model based on the linear economy, to one based on the principles of the circular economy and to set economic and environmental goals.

The circular economy (CE) concept, which has attracted the attention of business and policy makers as a new approach to sustainability (Ellen MacArthur Foundation, 2015; European Commission, 2018), is supported by underlying restoration, regeneration, and re-use of resources principles. Viewed from a management perspective, the essence of a circular business model is to exploit business opportunities in such a way that a company can create value not only economically but also environmentally (Pichlak & Szromek, 2022).

Within tourism, the adoption of CE principles still needs further dissemination between scholars and practitioners (Manniche et al., 2021; Rodríguez-Antón & Alonso-Almeida, 2019). The progress towards a CE requires a significant

---

B. Marco-Lajara (✉) · M. Úbeda-García · E. Poveda-Pareja · E. Manresa-Marhuenda
Department of Business Organisation, Universidad de Alicante, Alicante, Spain
e-mail: bartolome.marco@ua.es; mercedes.ubeda@ua.es; esther.poveda@ua.es; encarnacion.manresa@ua.es

© The Author(s) 2024
M. Segarra-Oña et al. (eds.), *Managing the Transition to a Circular Economy*, SpringerBriefs in Business, https://doi.org/10.1007/978-3-031-49689-9_9

transformation and many scholars emphasize the key role of green innovation in this regard (Sehnem et al., 2022; de Jesus & Mendonça, 2018).

In general, the innovations that are introduced in the tourism sector are fundamentally incremental (Lyons et al., 2007) and imply progressive modifications based on the reinforcement of existing knowledge and human interactions. These particularities about the nature of innovation in this sector suggest that the success in the formulation and implementation of green innovations depends to a large extent on human resource management and other people-driven factors. We propose that organizational culture plays a critical role in the successful implementation of circular business models and green innovations.

Advertising of green initiatives in the hospitality industry is abundant and growing but, nevertheless, there are few studies in this regard (Alonso-Almeida et al., 2016). Furthermore, evidence of the human side of these sustainability-oriented practices or their consequences is even more scarce with few exceptions (Sawe et al., 2021; Pham et al., 2023).

Therefore, to fill this gap in the literature on the opportunities of green innovation in the accommodation industry, we ask ourselves two questions: the first is whether green innovation allows hotel companies to improve their performance levels in economic and environmental terms; and the second, to what extent the existence of a robust organizational culture enhances the effect of green innovation on performance.

The remainder of this chapter is organized as follows. Section "Theory and Hypotheses" presents the theoretical background and hypotheses development, and then discusses the sample, data, and statistical procedures. Section "Data Analysis and Results" details the results of hypotheses testing, and the last section establishes the conclusions that are derived.

Theory and Hypotheses

Despite the multiplicity of terminologies (e.g. eco-innovation, sustainable innovation, environmental innovation, and others), there is a certain consensus in the literature towards the underlying meaning the concept, where all the definitions are usually focused on the improvements of environmental performance (Cai & Li, 2018). Green innovation in the accommodation industry refers to the development or modification of services, processes, organizational or marketing methods to contribute positively to the natural environment (intentionally or not) (Rennings, 2000). In fact, previous studies have considered that green innovation practices can positively and significantly affect the environmental performance of companies (Cai & Li, 2018; Asadi et al., 2020). In this sense, green innovation will improve and stimulate environmental performance because it allows adapting products and services to new sustainable demands, reducing emissions and the consumption of materials and energy, with a more efficient use of resources (Adegbile et al., 2017).

At once, one of the managers motivations for adopting green innovations is to help organizations attain a competitive advantage and achieve better economic performance (Font et al., 2017). Thanks to green innovation, it is possible to save on operating costs, improve the image of the company, comply with regulations and increase sales by serving new market segments (Quazi, 1999), in which they include consumers with a preference for sustainable products and more willing to pay a premium price for them. For these reasons, the effect of green innovation on economic performance has been studied in different research that have found a positive association between green innovation and performance.

The effect of green innovation in improving performance has also been the subject of analysis in the context of the tourism sector. Asadi et al. (2020) conducted research with a sample of 183 hotels in Malaysia and found that green innovation has a positive and significant influence not only on environmental performance, but also on economic performance. Accordingly, the following hypothesis is proposed:

H1: Green innovation affects corporate performance positively.

Tourism organizations are forced to continually reformulate their practices and processes as a consequence of the instability in which they carry out their activity (Nieves & Quintana, 2018). In this context, the employees are who promote and implement innovation (Lee & Hyun, 2016) since their direct interaction with the environment provides them with knowledge to identify areas that need improvement and can provide solutions to the problems of their company, their customers, and their environment (Karlsson & Skålén, 2015).

Organizational culture (OC) is defined as a set of beliefs and values shared by members of the same organization that influences the behavior of the company and its members as if they were unwritten rules (Cameron & Quinn, 1999). Consequently, OC is a powerful weapon that can be useful to promote creativity, risk taking, orientation towards results and the commitment of members to their company, reinforcing the organization's capacity to achieve its innovative goals due to the greater understanding of these by its members and their commitment to them. In other words, OC is inextricably linked to how individuals interact and behave (Luthra et al., 2017), so having a culture that encourages employees to develop new ways of working that promote innovation and circularity is a factor that can determine the survival of a company (Heyes et al., 2018).

OC was discussed by the scholars in numerous contexts considering certain issues related to the role of culture in the success of green innovation. In tourism sector, González-Rodríguez et al. (2019) have shown that the type of culture predominant in hotels determines the achievement in the implementation of sustainability innovations. In this line, the study by Cantele and Zardini (2018) shows that the impact of green innovation practices performance is conditioned by the organizational commitment of employees.

Therefore, it can be considered that a strong OC, in terms of employee commitment and innovation orientation, can enhance the relationship between green innovation and performance achieved, allowing us to formulate hypothesis 2:

H2: Robust organizational culture positively moderates the relationship between green innovation and corporate performance.

Research Methodology

To test the proposed relationships, we focus our research on tourism firms and, specifically, the population under study is made up of those tourist accommodation establishments in Spain (hotels, hostels, aparthotels and holiday complexes) that appear in the 'Alimarket' database. This pre-selection has been restricted by selecting those accommodations that are located exclusively in coastal municipalities because the innovations are characterized by responding to a specific business and market model, making it difficult to compare innovation between different competitive contexts and models, for instance between sun and beach tourism and urban tourism.

The rationale behind the choice of tourism as the target sector is based on its position as one of the main economic drivers at an international level, aspect that is justified by its contribution to global economy: in 2019 tourism contributed to 10.4% of world GDP and 10.6% of employment (Hosteltur, 2021). At the same time, from an environmental perspective, despite its dependence on the natural conditions of the tourist destination, the tourism sector and coastal tourism activity has traditionally been linked to negative externalities, such as the phenomenon of overtourism and specific problems such as landscape pollution (Gelbman, 2021). This dual reality of the tourism sector, in which economic and environmental pressures of the activity must be reconciled, make it an ideal context for our research.

Primary data obtained through a self-made questionnaire based on validated scales have been used. For its preparation, a detailed review of the literature and a pre-test were carried out with two professional experts and the executive directors of four companies.

In the final questionnaire used, items for measuring green innovation, organizational culture and corporate performance were included. Items were measured with 7-point Likert-scales where 1 meant strong disagreement and 7 meant strong agreement. Green innovation (GINN) is a first-order reflective construct that was measured with three items, based on the work of Xie and Zhu (2020). The organizational culture (OC) scale is intended to reflect the company's orientation towards teamwork, innovation, empowerment, and results-based formalization. This first-order reflective construct was measured with nine items adapted from the scale used by Oriade et al. (2021). Regarding the dependent variable corporate performance (PERF), a second-order reflective construct has been created, made up of one-dimensional constructs and made up of reflective items. The scales used for economic and environmental performance, of eight and five items, respectively, was adapted from Úbeda-García et al. (2021).

For data collection, an online version of the questionnaire (QualtricsXM software was used) was distributed among the CEOs of the hotel accommodations from September 2021 to January 2022.

Data Analysis and Results

In the current research, the authors' efforts have been directed to a specific model which is able to investigate the causal relationship between GINN and PERF and moderator role of OC. The relationships proposed in the model are tested using the Structural Equation Modeling (SEM) method based on analysis of variance: Partial Least Squares (PLS). For data treatment, Smart PLS 3 (version 3.3.9) was used.

The analysis of results is divided into two stages, following the theoretical recommendations, as it is a higher-order model of the hierarchical component approach (Lohmoller, 1989). In the first stage, the first-order results are analyzed, verifying that the first-order measurement model meets all theoretical requirements regarding reliability and construct validity. In addition, an assessment of the global first-order model has been carried out in which measures of approximate model fit have been applied (Henseler, 2017). This proves that the first-order global model has an adequate overall fit.

Once the criteria have been checked in the first-order stage and, to avoid repeating tables for the evaluation of the first- and second-order measurement model, the results obtained in the second-order model, that is, the advanced model that includes the theoretical hypotheses under study, are detailed in this research.

The evaluation of the PLS-SEM models focuses initially on the measurement models for each of the constructs. The aim is to assess the reliability and validity of the indicators of each construct. It is observed that the assessments of the measurement model improve with the creation of the second-order model.

To study the internal consistency, we analyzed three indicators of quality: Cronbach's α, rho_A, and composite reliability. To evaluate the convergent validity, we tested the average variance extracted (AVE), which shows that a group of indicators represent a single underlying construct (Henseler, 2017). See Table 9.1.

Finally, we evaluate the existence of discriminant validity to find to what extent a certain construct differs from the others. To do this, by applying the Fornell-Lacker criterion and the HTMT Inference criterion, we can verify that we have a good level of discriminant validity for the measurement model. In this sense, it has been verified by applying Fornell-Lacker and HTMT criterion that all the values are below 0.85

Table 9.1 Internal consistency reliability and convergent validity

	Cronbach's α	Rho_A	Composite Reliability	AVE
GINN	0.893	0.894	0.934	0.824
OC	0.918	0.936	0.932	0.606
PERF	0.705	0.744	0.869	0.769

Table 9.2 Multicollinearity assessment—VIF values

	GINN	OC	OCxGINN	PERF
GINN				
OC	0.502			
OCxGINN	0.110	0.263		
PERF	0.851	0.575	0.028	

Table 9.3 Path coefficients (direct and moderation effects)

	Path Coefficient	STDV	95% CI	Hypothesis
GINN → PERF	0.599***	0.060	[0.495, 0.601]	H1 Accepted
OCxGINN → PERF	0.094*	0.012	[0.012, 0.177]	H2 Accepted

***$p < 0.001$; **$p < 0.01$: *$p < 0.05$

(strict threshold proposed by Kline, 2011). Likewise, the HTMT Inference criterion is met, according to which the value 0.9 is not within the 95% CI (Gold et al., 2001).

After testing the reliability and the validity of the constructs, we analyzed the structural model. First, we checked for any problems related with collinearity, due to the need to avoid multicollinearity between the antecedent variables of each endogenous construct. Then we analyzed the path coefficients, the R^2 values and the effect sizes (f^2).

According to Hair et al. (2019), there are indications of collinearity when the variance inflation factor (VIF) is greater than 3. None of the VIF values obtained in this study are above the maximum value (Table 9.2).

After checking for collinearity, we evaluated the relevance of the relationships in the model. To this end we employed the PLS-SEM algorithm. To test whether the path coefficients are significant, we used bootstrapping. This technique allows us to test whether the hypothesized relationships are significantly different from 0 by analyzing the t statistic. Table 9.3 shows the significance levels for each relationship through their p values and their bootstrap confidence intervals.

The positive relationships proposed in Hypothesis 1 [GINN→PERF; $\beta = 0.599$; $p < 0.000$] are significant and have the proposed sign. We can confirm that green innovation has positive effects on corporate performance (that includes environmental and economic dimensions).

Additionally, the study assessed the moderating role of culture on the relationship between green innovation and performance. Without the inclusion of the moderating effect, the R-Sq value for PER was 0.482. This shows that 48.2% change in PERF is accounted by GINN. With the inclusion of the interaction term, the R-Sq increased to 0.518. This supposes an increase of 3.6% in variance explained in the dependent variable (PERF).

Further, significance of moderating effect was analyzed, the results revealed a positive and significant moderating impact of OC on the relationship between GINN and PERF ($\beta = 0.094$, $p < 0.05$), supporting H2. This implies that with an increase in role culture, the relationship between GINN and PERF is strengthened. Further, slope analysis is presented to better understand the nature of the moderating effects. As shown in Fig. 9.1, the line is much steeper for High OC, this implying that at

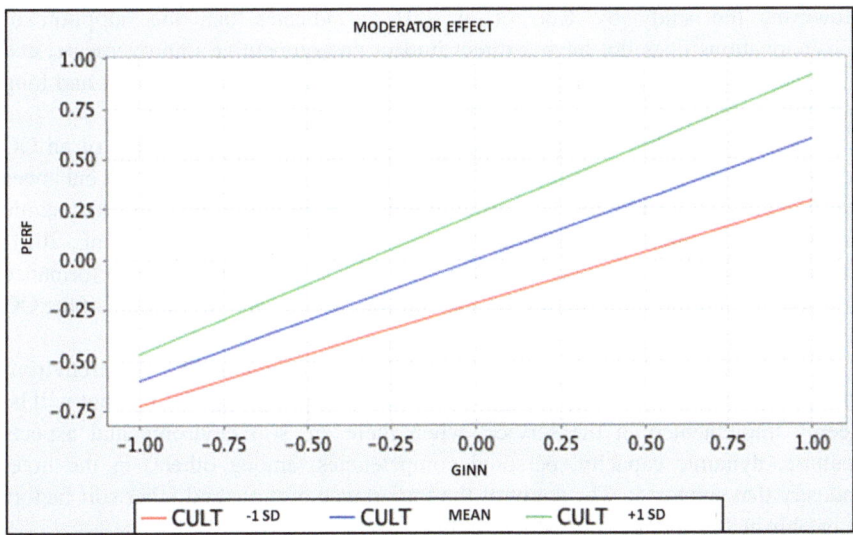

Fig. 9.1 Moderating effect

High Level of OC, the impact of green innovation on performance is much stronger in comparison with Low OC.

F-Square effect size was 0.018 and according to Cohen (1988) proposition, the contribution to explaining the endogenous construct (PERF) of the moderator effect is small.

Conclusions

The implementation of a circular economy model requires a long-term perspective and will depend on how companies create added value, the support offered by public authorities and policy makers, as well as how consumers perceive it. The viability of this transformation requires the active involvement of all the agents involved (Aboelmaged, 2018).

Currently, policy makers have already taken on the challenge of approving laws that encourage or force their implementation (for example, in the Balearic Islands-Spain) and, in tune with the evolution of user preferences and society in general, companies start to take small steps in the transition towards circular business models. However, progress is slow and costly because it imposes the need to make changes with the promise of economic and environmental gains whose materiality requires a medium and/or long-term time horizon.

Empirical research has shown that green innovation in hotels and other Spanish accommodation has a positive impact on corporate performance, coinciding with results obtained in other geographical contexts as disparate as China (Gu, 2023).

However, the study by Kuo et al. (2022) indicates that the adoption of eco-innovations does not have a direct impact on competitive improvements, and it is essential to have a competent human team committed to the complex and long process of change.

In this sense, it has been possible to demonstrate that the robustness of an OC focused on employee commitment and innovation is a positive factor that enhances the effect of GINN on PERF. This result highlights the importance of OC in dynamic and turbulent contexts and coincides with previous research (Oriade et al., 2021; Pham et al., 2023), by emphasizing that economic and environmental performance not only depends on the level of GINN adopted by hotels, but also strength of the OC that characterizes them.

In conclusion, green innovation in hotels is no longer limited to hard environmental aspects (construction, energy and consumption of raw materials), but will be deeply implemented in the service, when there are soft environmental aspects (culture, dynamic capacity, personal competencies, among others) in the hotel industry that support it. The study of the importance of these and other soft factors is established as a future line of research.

Finally, with a focus on practitioners, we would like to point out the importance of collaboration between stakeholders (policy makers, academics, firms, demand, etc.) in the development and distribution of knowledge on best practices as well as on technological solutions that promote the paradigm of the circular economy.

References

Aboelmaged, M. (2018). Direct and indirect effects of eco-innovation, environmental orientation and supplier collaboration on hotel performance: An empirical study. *Journal of Cleaner Production, 184*, 537–549.

Adegbile, A., Sarpong, D., & Meissner, D. (2017). Strategic foresight for innovation management: A review and research agenda. *International Journal of Innovation and Technology Management, 14*(04), 1750019.

Alonso-Almeida, M. M., Rocafort, A., & Borrajo, F. (2016). Shedding light on ecoinnovation in tourism: A critical analysis. *Sustainability, 8*(12), 1262.

Asadi, S., Pourhashemi, S. O., Nilashi, M., Abdullah, R., Samad, S., Yadegaridehkordi, E., Aljojo, N., & Razali, N. S. (2020). Investigating influence of green innovation on sustainability performance: A case on Malaysian hotel industry. *Journal of Cleaner Production, 258*, 120860.

Bruns-Smith, A., Choy, V., Chong, H., & Verma, R. (2015). Environmental sustainability in the hospitality industry: Best practices, guest participation, and customer satisfaction. *Cornell Hospitality Report, 15*(3), 6–16.

Cai, W., & Li, G. (2018). The drivers of eco-innovation and its impact on performance: Evidence from China. *Journal of Cleaner Production, 176*, 110–118.

Cameron, K. S., & Quinn, R. E. (1999). *Diagnosing and changing organizational culture*. Addison-Wesley.

Cantele, S., & Zardini, A. (2018). Is sustainability a competitive advantage for small businesses? An empirical analysis of possible mediators in the sustainability–financial performance relationship. *Journal of Cleaner Production, 182*, 166–176.

Cohen, J. (1988). *Statistical power analysis for the behavioral sciences* (2nd ed.). Lawrence Erlbaum.

de Jesus, A., & Mendonça, S. (2018). Lost in transition? Drivers and barriers in the eco-innovation road to the circular economy. *Ecological Economics, 145,* 75–89.

Ellen Macarthur Foundation. (2015). *Global partners.* https://www.ellenmacarthurfoundation.org/about/global-partners.

European Commission. (2018). *Circular economy: Implementation of the circular economy action plan.* http://ec.europa.eu/environment/circular-economy/index_en.htm.

Font, X., Elgammal, I., & Lamond, I. (2017). Greenhushing: The deliberate under communicating of sustainability practices by tourism businesses. *Journal of Sustainable Tourism, 25,* 1007–1023.

Gelbman, A. (2021). Tourist experience and innovative hospitality management in different cities. *Sustainability, 13*(12), 6578.

Gold, A. H., Malhotra, A., & Segars, A. H. (2001). Knowledge management: An organizational capabilities perspective. *Journal of Management Information Systems, 18*(1), 185–214.

González-Rodríguez, M. R., Martín-Samper, R. C., Köseoglu, M. A., & Okumus, F. (2019). Hotels' corporate social responsibility practices, organizational culture, firm reputation, and performance. *Journal of Sustainable Tourism, 27*(3), 398–419.

Gössling, S., Scott, D., & Hall, C. M. (2013). Challenges of tourism in a low-carbon economy. *Wiley Interdisciplinary Reviews: Climate Change, 4*(6), 525–538.

Gu, S. (2023). Green innovation; a way to enhance economic performance of Chinese hotels. *International Journal of Innovation Science, 15*(3), 406–426.

Hair, J., Hult, G., Ringle, C., Sarstedt, M., Castillo-Apraiz, J., Cepeda Carrion, G., & Roldán, J. (2019). *Manual de Partial Least Squares Structural Equation Modeling (PLS-SEM)* (2nd ed.). OmniaScience Scholar.

Hall, C. M. (2019). Constructing sustainable tourism development: The 2030 agenda and the managerial ecology of sustainable tourism. *Journal of Sustainable Tourism, 27*(7), 1044–1060.

Henseler, J. (2017). Bridging design and behavioral research with variance-based structural equation modeling. *Journal of Advertising, 46*(1), 178–192.

Heyes, G., Sharmina, M., Mendoza, J. M. F., Gallego-Schmid, A., & Azapagic, A. (2018). Developing and implementing circular economy business models in service-oriented technology companies. *Journal of Cleaner Production, 177,* 621–632.

Higgins-Desbiolles, F., Carnicelli, S., Krolikowski, C., Wijesinghe, G., & Boluk, K. (2019). Degrowing tourism: Rethinking tourism. *Journal of Sustainable Tourism, 27*(12), 1926–1944.

Hosteltur. (2021). *La aportación del turismo al PIB mundial cae a la mitad por la pandemia.* https://www.hosteltur.com/143169_la-aportacion-del-turismo-al-pib-mundial-cae-a-la-mitad-por-la-pandemia.html

Karlsson, J., & Skålén, P. (2015). Exploring front-line employee contributions to service innovation. *European Journal of Marketing, 49*(9/10), 1346–1365.

Kline, R. (2011). *Principles and practice of structural equation modeling.* New York.

Koçak, E., Ulucak, R., & Ulucak, Z. Ş. (2020). The impact of tourism developments on CO2 emissions: An advanced panel data estimation. *Tourism Management Perspectives, 33,* 100611.

Kuo, F. I., Fang, W. T., & Lepage, B. A. (2022). Proactive environmental strategies in the hotel industry: Eco-innovation, green competitive advantage, and green core competence. *Journal of Sustainable Tourism, 30,* 1240–1261.

Lee, K. H., & Hyun, S. S. (2016). An extended model of employees' service innovation behavior in the airline industry. *International Journal of Contemporary Hospitality Management, 28*(8), 1622–1648.

Lohmoller, J. B. (1989). *Latent variables path modeling with partial least squares.* Physica.

Luthra, S., Govindan, K., Kannan, D., Mangla, S. K., & Garg, C. P. (2017). An integrated framework for sustainable supplier selection and evaluation in supply chains. *Journal of Cleaner Production, 140,* 1686–1698.

Lyons, R. K., Chatman, J. A., & Joyce, C. K. (2007). Innovation in services: Corporate culture and investment banking. *California Management Review, 50*(1), 174–191.

Manniche, J., Larsen, K. T., & Broegaard, R. B. (2021). The circular economy in tourism: Transition perspectives for business and research. *Scandinavian Journal of Hospitality and Tourism, 21*(3), 247–264.

Nieves, J., & Quintana, A. (2018). Human resource practices and innovation in the hotel industry: The mediating role of human capital. *Tourism and Hospitality Research, 18*(1), 72–83.

Oriade, A., Osinaike, A., Aduhene, K., & Wang, Y. (2021). Sustainability awareness, management practices and organisational culture in hotels: Evidence from developing countries. *International Journal of Hospitality Management, 92*, 102699.

Pham, N. T., Chiappetta Jabbour, C. J., Vo-Thanh, T., Huynh, T. L. D., & Santos, C. (2023). Greening hotels: Does motivating hotel employees promote in-role green performance? The role of culture. *Journal of Sustainable Tourism, 31*(4), 951–970.

Pichlak, M., & Szromek, A. R. (2022). Linking eco-innovation and circular economy—A conceptual approach. *Journal of Open Innovation: Technology, Market, and Complexity, 8*(3), 121.

Quazi, H. A. (1999). Implementation of an environmental management system: The experience of companies operating in Singapore. *Industrial Management & Data Systems, 99*(7), 302–311.

Rennings, K. (2000). Redefining innovation—eco-innovation research and the contribution from ecological economics. *Ecological Economics, 32*(2), 319–332.

Rodríguez-Antón, J. M., & Alonso-Almeida, M. D. M. (2019). The circular economy strategy in hospitality: A multicase approach. *Sustainability, 11*(20), 5665.

Sawe, F. B., Kumar, A., Garza-Reyes, J. A., & Agrawal, R. (2021). Assessing people-driven factors for circular economy practices in small and medium-sized enterprise supply chains: Business strategies and environmental perspectives. *Business Strategy and the Environment, 30*(7), 2951–2965.

Sehnem, S., de Queiroz, A. A. F. S., Pereira, S. C. F., dos Santos Correia, G., & Kuzma, E. (2022). Circular economy and innovation: A look from the perspective of organizational capabilities. *Business Strategy and the Environment, 31*(1), 236–250.

Sharpley, R. (2020). Tourism, sustainable development and the theoretical divide: 20 years on. *Journal of Sustainable Tourism, 28*(11), 1932–1946.

Úbeda-García, M., Claver-Cortés, E., Marco-Lajara, B., & Zaragoza-Sáez, P. (2021). Corporate social responsibility and firm performance in the hotel industry. The mediating role of green human resource management and environmental outcomes. *Journal of Business Research, 123*, 57–69.

Xie, X., & Zhu, Q. (2020). Exploring an innovative pivot: How green training can spur corporate sustainability performance. *Business Strategy and the Environment, 29*(6), 2432–2449.

Open Access This chapter is licensed under the terms of the Creative Commons Attribution 4.0 International License (http://creativecommons.org/licenses/by/4.0/), which permits use, sharing, adaptation, distribution and reproduction in any medium or format, as long as you give appropriate credit to the original author(s) and the source, provide a link to the Creative Commons license and indicate if changes were made.

The images or other third party material in this chapter are included in the chapter's Creative Commons license, unless indicated otherwise in a credit line to the material. If material is not included in the chapter's Creative Commons license and your intended use is not permitted by statutory regulation or exceeds the permitted use, you will need to obtain permission directly from the copyright holder.

Chapter 10
Circular Economy Self-assessment Tool for Hotels

Marival Segarra-Oña, Ángel Peiró-Signes, Joaquín Sánchez-Planelles, and Esther Poveda-Pareja

Introduction

The services sector has been rapidly growing over the last decades and nowadays represents one of the economic engines of many countries around the world. In particular, tourism industry has contributed significantly to the GDP of numerous countries and has been on the rise (UNWTO, 2022). Unlike other sectors whose impacts and interrelationships are more delimited, tourism is strongly associated with a range of services, from agriculture and finance to construction and retail (Sorin and Einarsson, 2020). Indeed, these interrelations with other industries magnifies the impact of this industry on the economy, both in terms of economic activity and job creation (WTTC, 2019). However, this industry is also known to provoke negative externalities on the environment, leading to its degradation and a rise in Green House Gas (GHG) emissions (Lenzen et al., 2018), among other harmful effects. Particularly, the transportation and the hospitality industry the tourism activities with the most important impact on CO_2 emissions (UNTWO, 2008) and global trends in both sectors continue with a steady growth. Indeed, several studies have already reported a high use of resources, such as water or energy, and a high production of waste (Girard and Nocca, 2017; Alonso-Almeida, 2012; Manniche et al., 2017). Then, despite its positive impacts to the economies, it is important to take into account the negative environmental consequences of the industry in order to be able to achieve a balance between economic growth and environmental care.

M. Segarra-Oña (✉) · Á. Peiró-Signes · J. Sánchez-Planelles
Department of Business Organisation, Universitat Politècnica de València, Valencia, Spain
e-mail: maseo@omp.upv.es; anpeisig@omp.upv.es

E. Poveda-Pareja
Department of Business Organisation, Universidad de Alicante, Alicante, Spain
e-mail: esther.poveda@ua.es

To ensure the welfare and proper living standards of current and future generations, it is imperative that the travel and tourism sector develops a strategic paradigm shift, where it rethinks its values, purpose, business models, and value chains, what should benefit from its long-term viability and sustainability (UN, 2016). The adoption of this sustainable path can be provided by the Circular Economy (CE). CE allows for a better use of inputs and a better management of the outputs to pursue eliminating, reducing, reusing, recycling or recovering materials in the business operation (Kirchherr et al., 2017). So far, the concept of CE has been mainly addressed in resource-intensive industries, however, the above highlighted interrelations among industries suggest that CE requires a holistic view of the business and the active implication of the stakeholders and surrounding industries (Marino and Pariso's, 2021).

In this line, the European Union is taking steps to reach the climate neutrality by 2050 based on CE. Initiatives such as the New Circular Economy Action Plan (COM 98, 2020) or the European Green Deal (COM 640, 2019) focus the action on extending the use of resources and on minimizing the amount of waste generated. Considering these two main factors, to move forward to a circular model, organizations require indicators and methods to be able to measure and evaluate their status and progress with respect to CE. There have been several approaches when measuring CE, such as "A New Industrial Strategy for Europe" (COM 102, 2020), Longevity indicator, Resource Potential Indicator, Sustainable Process Index, Input Output Analysis, Material Flow Analysis or Life Cycle Assessment, among others. However, many of these indicators cannot be applied to organizations and require some of their own ones (Vinante et al., 2021). Consequently, there is no consensus when evaluating CE strategies and plans. Thus, organizations' evolution towards a CE needs instruments that allow them to evaluate and communicate the results of their transition to a CE, what requires indicators, measurement methods and procedures to be standardized.

Even though the hotel industry has been adopting eco-friendly practices since the beginning of the twenty-first century (Alvarez-Gil et al. 2001) and, more recently, the world's leading hotel companies have already adopted some CE principles and practices, many hotels are still lagging behind in the race towards sustainability. Moreover, there has been little scientific research on why more quantity of hotels are not yet implementing CE (Rodriguez et al., 2020) or how hotels can adopt CE values. Particularly, there is a lack of research in which tools should be applied for CE measuring and evaluation in the hospitality industry, and little amount of good practices and lessons have been learned from CE implementations.

Taking into account this context, the main goal of this chapter is to develop a basic measurement and evaluation framework for measuring and evaluating the circular level of hotels as a tool for monitoring and planning their circular actions. This, eventually, will lead hotels to implement good practices leading to an increase of their circularity level.

The structure of this chapter is as follows. Section "Literature Review" is focused on the theoretical framework. Section "Methodology" presents the methodology of the study. Section "The Self-assessment Tool" along describes the indicators and the evaluation framework. Finally, the conclusions and further lines of development of the self-evaluation tool are examined in section "Conclusions".

Literature Review

CE is a concept that goes further than reduction, reuse and recycling (Ghisellini et al., 2015), extending the nature of the intervention from the conception of the product or service, through redesigning, to the extension of the life of the resource, through recovering, or the product, through remanufacturing. The hospitality industry actions towards CE have been reported mainly in energy, water and waste recycling. In fact, recent legislation in some pioneer destinations, such as the Balearic Islands (Comunidad Autónoma de las Illes Balears, 2022), focusses in these areas. Since, hotel industry is an operational service industry, most of the CE implementations have been directed to improve the circularity of the hotel operations. Reduction strategies are the most common approaches to introduce sustainable and circular practices in hotels (Manniche et al., 2017). Additionally, eco-innovations (Florido et al., 2019; Alonso-Almeida et al., 2016) are popular actions in this area. The best practices in energy, water and waste reduction are related to the use, for example, of renewable energies or efficient lightening (Vourdaubas, 2016; Girard and Nocca, 2017), the reduction of water consumption with water aerators, environmental responsible laundry services or rain or graywater water storage (Alonso Almeida et al., 2017; Fernandez-Robin et al., 2019; Manniche et al., 2017), the reduction of waste through recovery, reuse or valorization of the disposed materials (Florido et al., 2019; Girard and Nocca, 2017), or through better design, planning and management of food services to reduce leftovers and food waste. In most cases, still, there are no strategies to close the material's cycles and tend to zero waste, which requires collaboration and symbiotic relations between the hotel industry and the near-by industrial environment (Singh and Giacosa, 2019).

Among the different CE assessment indicators and methods we can highlight the Longevity indicator, the Resource Potential Indicator, the Sustainable Process Index, the Input Output Analysis, the Material Flow Analysis or the Life Cycle Assessment. However, many of the studies and indicators are focused on and developed for a product or territorial level, with a few exceptions that have been developed to evaluate circularity at a company level (Vinante et al., 2021), such as the CE performance indicators created by EU to track organizations' progress toward the CE (A New Industrial Strategy for Europe, COM 102, 2020) or CE indicators and methods for organizations reported by some authors (Kravchenko et al., 2019; Moraga et al., 2019; Parchomenko et al., 2019; Saidani et al., 2017; Kristensen and Mosgaard, 2020; Lindgreen et al., 2020; Vinante et al., 2021; Franco et al., 2021). From these studies, we can conclude that there is a lack of consensus when evaluating CE because of the quantity and variety of indicators, methods and approaches available. Somehow, one size-fits-all tool for CE measurement is not a viable option. The differences between the type of product and services that they offer, the operations, the supply chains and the connections with other industries suggest that specific industry indicators should be developed to better orient companies' actions towards CE. Industries and organizations within these industries require their own CE indicators as well as tools to evaluate the degree of CE implementation, which make it simpler to calculate these indicators. This will ease the assessment and the movement towards the CE. Additionally, it will allow organizations to benchmark, market and develop competitive advantages of their progress with respect to their competitors.

Methodology

In this chapter, we propose a basic tool to evaluate the level of circularity in the hotels. This tool is intended to be complemented in the future as hotels take steps towards circularity with one or two more levels. We propose to walk in these levels towards a more detailed and extended model of indicators and to a more detailed number of best practices. Additionally, we suggest that the evaluation based on the indicators should progressively be more centered in the results and less on the actions taken.

The model builds on the existing measurement framework of the Circularity regulations approved by the Balearic Islands for the hotel industry, and proposes extensions to overcome one of the main issues that the regulation lacks of. The proposed indicators in the regulation are scarce and in many cases are not contextualized. Then, it is difficult to use it to benchmark hotels with similar characteristics and are indicators that are very vague to use it as an incentive for actions. That is, hoteliers cannot catch a glimpse of all the good practices that are aligned with CE and that can be implemented with the current indicators. Therefore, we propose a set of good practices indicators to orient the action plan. In other words, by evaluating the circular good practices of the hotel we are, on one hand, informing hoteliers about a set of actions to be taken, and, on the other hand, stablishing the bases for a circularity action plan. The tool is accompanied of a best practices questionnaire to guide the hotel in their journey to the creation of their circularity plan, including the measurement, evaluation and planification of the actions. In addition, the number of possible indicators and best practices are enormous and the InnoEcoTur project aims at facilitating the transition to CE to hotels, which in many cases are SMEs and lack of resources and specific CE knowledge. Then, after evaluating the situation of the industry in several Living labs (InnoEcoTur 2022, 2023), we decided to promote in this first approach as a basic self-assessment tool.

This tool aims to be simple enough for hoteliers to understand, retrieve and calculate the data needed for the indicators and to motivate them to commit to CE. Additionally, we have designed so it can evolve to more and more detailed indicators. For example, related best practices in this initial tool have been grouped and are evaluated together, so in future developments of the tool (i.e. intermediate or advance self-evaluation tool) they can be configured as independent questions, and therefore, evaluate the progression of the hotel in CE at a higher detail. Additionally, scales have been established in the tool considering actual industry data and the tool weights mainly the actions taken over the results. Then, we expect a hotel that has been working in sustainability to perform well in the tool now, and therefore, encouraging them rapidly to evolve other levels of the tool and, therefore, to higher level of circularity. On the other side, a hotel that traditionally hadn't been engaged in many sustainable initiatives will be able to orient their actions implementing the suggested good practices and start rising their evaluation results with simple measures, maintaining their motivation to improve their circularity in these early stages.

The tool is divided into 5 dimensions: Circularity management, Energy, Water, Waste and Food waste.

We developed the tool in following a step procedure:

Step 1. Exploration

- Identify regulations with mandatory indicators.
- Identify existing sustainability indicators that are related to circular economy. Tools and guidelines (CTI Tool, 2020; CEEI, 2020; CircularTRANS, 2020; CIRCelligence, 2020; Circulytics, 2020, among others), best practices and RSC related reports in the industry (i.e. Iberostar).
- Identify existing indicators for circular economy in other industries (Kravchenko et al., 2019; Moraga et al., 2019; Parchomenko et al., 2019; Saidani et al., 2017; Kristensen and Mosgaard, 2020; Lindgreen et al., 2020; Vinante et al., 2021; Franco et al., 2021).

Step 2. Analysis

- Compile indicators and categorize them according to the initial dimensions and the utility (utility measured as how informative is the indicator about the level of circularity of the hotel and how easy is this indicator to obtain and monitor).
- Select hospitality measures and characteristic useful for comparison purposes (i.e. ADR, occ, revPAR, category, management type, location, etc.) and preselect complementary indicators to build benchmarkable EC indicators.
- Group the related indicators attending to the purpose and dimension.

Step 3: Filtering and selection

- Filter indicators eliminating redundancies.
- Separate quantitative indicators and best practices related indicators.
- Adapt general indicators to industry specifics to be able to compare between hotels. (i.e. indicator per number of occupied rooms, indicator per number of guests, indicator per square meter,...).

Step 4: Exploration and selection of scales and levels for quantitative indicators

- Identify existing scales or value levels for potential indicators (Greenview, AVEN,....)
- Eliminate or convert scales from subjective to objective scale.

- Compare and homogenize indicator scales and levels.
- Propose level thresholds based on existing information from trusted sources.
- Standardize scale intervals according to 4 levels of EC (low, medium, high and very high).

Step 5: Exploration and selection of the best practices

- Identify existing best practices.
- Group them according to their relation (i.e. best practices to reduce water consumption in the room, best practices to reduce water consumption in related services (laundry,...).
- Elaborate question.

Step 6: Distribution and weighting of the indicators.

- Distribute mandatory indicators in the dimensions. Complete with the relevant quantitative indicators and include dimension related best practices questions.
- Assign percentages to dimensions and to indicators within each dimension.

Step 7: Validation of the tool by experts and hotels (in-progress).

The Self-assessment Tool

To determine the dimensions, we started from the priority areas that are established in Balearic Islands Law on urgent measures for the sustainability and circularity of tourism in the Balearic Islands: Energy, Water, Waste and Food Waste. Additionally, we incorporated a dimension on the Circular planning and evaluation (see Fig. 10.1).

Each dimension is structured in two parts, indicators and good practices. Both indicators and good practices include the minimums established by Law 3/2022 of the Balearic Islands, and incorporates additional indicators that allow a first circularity assessment of tourist accommodation to be made. For example, Circular Planning and Evaluation includes the list of tasks and actions, periodicity, distribution of resources, investments, protocols and any other human, material and economic means necessary to comply with the circularity plan. In the Energy dimension, the hotels have to report indicators such as the degree of energy efficiency of the building or the installations, the carbon footprint, the percentage of energy coming from renewable sources or the energy storage capacity. Water dimension analyzes, on the one hand, the consumption and self-supply capacity from water resources, and, on the other, the strategy, management planning, facilities and associated equipment related to water consumption. Waste is mainly to increase the correct separation of the different waste fractions that can be properly recycled and, therefore, to reduce the unsorted waste. Finally, Food Waste evaluates incoming food, considering both the proximity and the sustainable origin of the food supply, the food planning and the strategies to reduce leftovers.

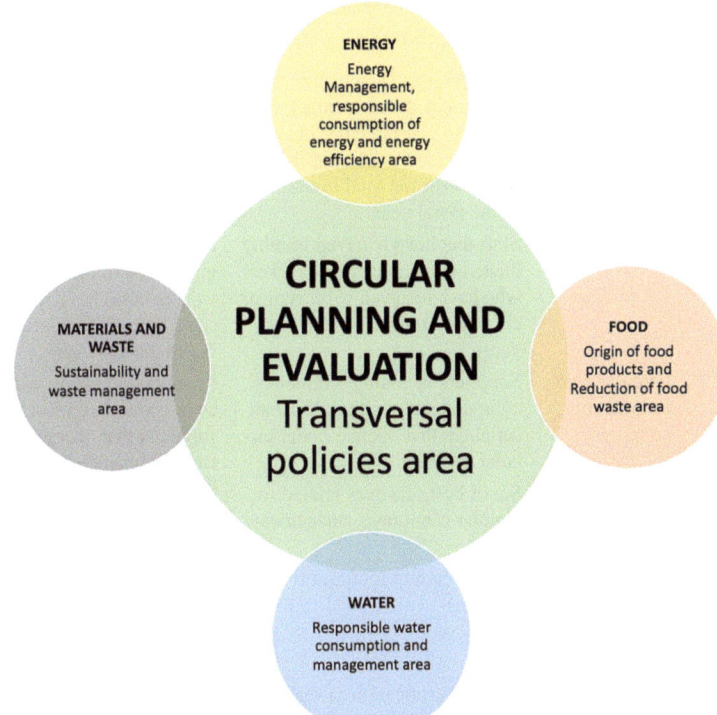

Fig. 10.1 Dimensions of the CE self-assessment tool for hotels

Each dimension is divided into two blocks, the data entry part needed to calculate the indicators and the best practices' part. Table 10.1 summarizes the data provided in each of the 5 dimensions, subdimension and the related indicators used in the evaluation framework.

Data is combined with hotel's operational data, such as number of room-nights, number of guest-nights or square meters of the hotel, to obtain the different indicators.

The questionnaire developed by Innoecotur (https://innoecotur.webs.upv.es) includes quantitative data that has to be retrieved by hotel managers and yes/no question related to sets of best practices.

The sources of indicator levels where retrieved based on existing data from public and private organizations working and both hospitality and sustainability indicators. We used Cornell Hotel Sustainability Benchmarking (CHSB) Index (Ricaurte and Jagarajan, 2021) developed by Greenview, Energy distribution for hotels from PricewaterhouseCoopers report (2013), Waste management and circular economy report from EAE Business School (Seguí et al., 2018), Best Environmental Management Practice in the Tourism Sector report (Styles et al., 2013), Manual of good practices for improving the energy efficiency of hotels in the Canary Islands

Table 10.1 Dimensions, subdimension and the related indicators of the CE self-assessment tool for hotels

Dimension	Subdimension	Data	Indicator
Circular planning and evaluation	Investment	Fraction of the investment linked to good circular practices1 made by the establishment (% of the total investment of the last 3 years)	Fraction of the investment linked to good circular practices1 made by the establishment (% of the total investment of the last 3 years)
	Training	Staff that have received training related to circularity (% workers who have received some type of circularity training in the last year)	Staff that have received training related to circularity (% workers who have received some type of circularity training in the last year)
	Circular commitment of suppliers	Suppliers operating with a circular code of conduct (% of all suppliers that operate under circular practices)	Suppliers operating with a circular code of conduct (% of all suppliers that operate under circular practices)
	Best practices	Set of best practices related to circular economy management	Number of best practices over the total set related to circular economy management
Energy	Carbon footprint	Annual carbon footprint (t CO_2 eq./year)	Carbon footprint per occupied room (t CO_2 eq./room)
			Carbon footprint per square meter (t CO_2 eq./m^2)
	Energy certifications	Energy certification of the building (grade A to G)	Energy certification of the building (grade A to G)
	Self-supply of energy	Self-supply power capacity (% self-generated/self-consumed energy from renewable sources with respect to total energy consumed per year in the establishment)	Self-supply power capacity (% self-generated/self-consumed energy renewable sources with respect to total energy consumed per year in the establishment)
		Installed renewable power (kWh)	Installed renewable power (kWh)
		Storage capacity (kWh)	Storage capacity (kWh)
	Energy parameters	Total annual energy consumption of the accommodation (MWh/year)	Energy consumption per occupied room (kwh per occupied room)
		Annual consumption of thermal energy of the accommodation (MWh/year)	Energy consumption per square meter (kwh per square meter)
		Annual electricity consumption of the accommodation (MWh/year)	
	Electric instalations efficiency	Lighting systems (grade A to G) Appliances and kitchen facilities (grade A to G) Appliances and room facilities (grade A to G) Elevators (grade A to G)	Average value of Energy certification of the installations (grade A to G)

(continued)

Table 10.1 (continued)

Dimension	Subdimension	Data	Indicator
	Best practices	Set of best practices related to energy management	Number of best practices over the total set related to energy management
Water	Self-supply capacity	Water self-supply capacity (% volume of self-withdrawal and/or treated water with respect to the total water in the establishment and in its facilities)	Water self-supply capacity (% volume of self-withdrawal and/or treated water with respect to the total water in the establishment and in its facilities)
	Water consumption	Annual water consumption (from the public network) (L/year)	Water consumption per occupied room (liters per occupied room) Water consumption per square meter (liters per square meter)
	Best practices	Set of best practices related to water management	Number of best practices over the total set related to water management
Waste	Waste hierarchy	Selective waste collection (Estimated volume, per overnight stay, of waste collected selectively (sum of fractions of paper and cardboard, glass, containers, among others)) (m3/night)	Percentage of waste collected selectively (%)
		Recycling of waste from works, reforms and demolitions (% of recycled waste from works, reforms and demolitions with respect to the total waste from constructions and demolitions generated during the last financial year, or, failing that, during the last construction project) (%)	Percentage of construction waste recycled (%)
	Generated waste	– Organic waste – Plastic waste – Waste paper and cardboard – Glass waste – Other waste collected selectively – Waste not deposited selectively – Total waste (m3)	
	Best practices	Set of best practices related to waste management	Number of best practices over the total set related to waste management
Food waste	Sustainable food expenditure	Consumption of zero-kilometer products (% of the	Expenditure of zero-kilometer products (%)

(continued)

Table 10.1 (continued)

Dimension	Subdimension	Data	Indicator
		establishment's total expenditure on food and beverages)	
		Shopping basket that minimizes the use of packaging (% represented by products in bulk and/or with reusable/biodegradable containers over the total expense items for supplies of the establishment)	Expenditure of products that minimizes packaging (%)
	Best practices	Set of best practices related to food waste management	Number of best practices over the total set related to food waste management

(CEHAT, 2008), World Tourism Organization (UNWTO, 2022) report or the Valencian Energy Agency information on hotels (AVEN, 2003), among others. The sources provided different kind of information for the proposed indicators, some of the data was related to real consumption (i.e. CHSB data), while others were related to best practices values (i.e. European commission report). We avoided those indicators related to overall values because they do not consider the size effect or the occupancy of the hotel. Therefore, we focused on those where the unit of consumption of the indicator was measured per available room, per guest or per square meter. In a few cases, values were related to the category of the hotel or the size in terms of number of rooms. For these cases we considered an average value as a reference.

Quantitative data, such as carbon footprint have to be considered in a way that results from different hotels can be compared and considering the operational characteristics of the hotel. For example, although the Law 3/2022 requires the calculation of the annual carbon footprint per overnight stay, this indicator does not consider the size of the hotel. Therefore, we included in this type of indicators the evaluation of the indicator per square meter to overcome the size effect.

Some of the key measures, such as the energy or water consumption, were extracted from The Cornell Hotel Sustainability Benchmarking Index (Ricaurte and Jagarajan, 2021). The CHSB index is an indicator that benchmark hotels around the world in different sustainable indicators such as carbon footprint, energy or water. It covers more than 25,000 hotels from different geographical regions. The index indicators provide a good base value to consider hotel performance for energy and water dimensions. Particularly, for the Mediterranean area we were able to retrieve data from 600+ hotels belonging to Mediterranean costal area and 500+ to Mediterranean interior area. Data is provided in quartile values for indicators. We used quartile weighed average values for the two regions (1100+ hotels) as threshold values for our scales. Additionally, CHBS already includes industry related base to the indicators as it gives information per occupied room and per square meter. We complemented the information with other sources such as the Best Environmental

Management Practice in the Tourism Sector report (Styles et al., 2013) or the guide to saving and energy efficiency in hotel establishments in the Valencian Community (AVEN, 2003). In the first one, we found several threshold values for best practices that are aligned with Circular indicators. For example, Hamele and Eckardt (2006) state that in European hotels, every guest consumes on average around 394 liters per night, while a best practice for water consumption is reported to be an average water consumption less than 200 liters/tourist and day, corresponding less than 140 l/guest-night for accommodations (Styles et al., 2013). The second document indicates that an excellent hotel should have Energy consumption lower than 240–365 kw/m2-year and Water consumption lower than 120–220 l/guest-year depending on the size, measured by the number of rooms. These and other sources, such as Manual of good practices for improving the energy efficiency of hotels in the Canary Islands (CEHAT, 2008) or Waste management and circular economy (Seguí et al., 2018) have been used to calibrate the indicator levels of the different indicators.

Energy certification indicators are defined as a letter scale from A (the most efficient) to G (the least efficient). We converted the letter scale to a numerical scale from 1 ("A" certifications) to 0 ("G" certifications) and evaluated the hotel according to the numerical scale.

Percentage indicators reported in the questionnaire are considered and evaluated attending to a reasonable maximum value. For example, for the percentage of energy coming from renewable energies, the maximum value that we found in a hotel is 38%. Taking into account that all the energy would be difficult to come from renewable sources, we set a maximum value in 30% and distributed the evaluating points accordingly. Thus, a hotel with 15% of the energy from renewable sources will get 50% of the points in this part. In the cases where high percentage values can be achieved, i.e. amount of personnel with CE training, we maintained the scale from 0 to 100% and the corresponding percentage of the points were assigned.

Best practices in each question are evaluated as a set and hotels either have implemented any of these best practices or not. Best practices have been also grouped in each dimension attending to the area of the dimension which they are related to, i.e. water saving best practices related to the hotel room. Therefore, for each dimension the questionnaire asks for different sets of best practices. The evaluation of best practices is done by evaluating the positive responses overall. Thus, a hotel performing at least one of the best practices in a specific area (one question in the questionnaire) will get full credit for that question. This has been designed for different purposes. First, we want to motivate hotels that have been working on sustainability highlighting the previous effort done. Second, these allows the tool to promote other best practices that hotels haven't implemented yet and that can be included in the circularity plan. Third, this will allow more advanced levels of the tool to disaggregate the best practices into single questions, giving credit only on those best practices that have been implemented and, therefore, pushing hotels to advance and implement further actions towards circularity.

Table 10.2 summarizes the evaluation scheme. We distributed one hundred points into the 5 dimensions, 20 each, and assigned a specific percentage of these points to each of the subdimensions attending to the importance of these indicators in the

Table 10.2 Scales and score distribution for dimensions and subdimensions of the CE self-assessment tool for hotels

Dimension	Subdimension	Scale (min—thresholds—max)	Indicator
Circular planning and evaluation 20%	Investment 20%	0%—10%—20%—30%—40%	Fraction of the investment linked to good circular practices1 made by the establishment (% of the total investment of the last 3 years)
	Training 5%	0%—25%—50%—75%—100%	Staff that have received training related to circularity (% workers who have received some type of circularity training in the last year)
	Circular commitment of suppliers 5%	0%—25%—50%—75%—100%	Suppliers operating with a circular code of conduct (% of all suppliers that operate under circular practices)
	Best practices 70%	0—4,75—9,5—14,25—19	Number of best practices over the total set related to circular economy management
Energy 20%	Carbon footprint 15% (10% + 5%)	24,5—19—13,5—8—2,5 (t CO2 eq. /occupied room) 85—70—55—40—25 (t CO2 eq. /square meter)	Carbon footprint per occupied room (t CO2 eq./room) Carbon footprint per square meter (t CO2 eq./m2)
	Energy certifications 10%	A—B—C—D—E—F—G	Energy certification of the building (grade A to G)
	Self-supply of energy 10%	0%—7,5%—15%—22,5%—30%	Self-supply power capacity (% self-generated/self-consumed energy renewable sources with respect to total energy consumed per year in the establishment)
			Installed renewable power (kWh)
			Storage capacity (kWh)
	Energy parameters 15% (10% + 5%)	87,5—70—52,5—35—17,5 (kwh per occupied room) 330—280—230—180—130 (kwh per square meter)	Energy consumption per occupied room (kwh per occupied room) Energy consumption per square meter (kwh per square meter)
	Electric instalations efficiency 10%	Lighting systems A—B—C—D—E—F—G Appliances and kitchen facilities A—B—C—D—E—F—G Appliances and room facilities A—B—C—D—E—F—G Elevators A—B—C—D—E—F—G	Average value of Energy certification of the installations (grade A to G)

(continued)

Table 10.2 (continued)

Dimension	Subdimension	Scale (min—thresholds—max)	Indicator
	Best practices 40%	0—2—4—6—8	Number of best practices over the total set related to energy management
Water 20%	Self-supply capacity 20%	0%—25%—50%—75%—100%	Water self-supply capacity (% volume of self-withdrawal and/or treated water with respect to the total water in the establishment and in its facilities)
	Water consumption 40% (30% + 10%)	685—570—455—340—225 (liters per occupied room) 3000—2650—2000—1400—800 (liters per square meter)	Water consumption per occupied room (liters per occupied room) Water consumption per square meter (liters per square meter)
	Best practices 40%	0—1,5—3—4,5—6	Number of best practices over the total set related to water management
Waste 20%	Waste hierarchy 60% (50% + 10%)	0%—25%—50%—75%—100%	Percentage of waste collected selectively (%)
		0%—25%—50%—75%—100%	Percentage of construction waste recycled (%)
	Generated waste		
	Best practices 40%	0—3—6—9—12	Number of best practices over the total set related to waste management
Food waste 20%	Sustainable food expenditure 40% (20% + 20%)	0%—12,5%—25%—37,5%—50%	Expenditure of zero-kilometer products (%)
		0%—12,5%—25%—37,5%—50%	Expenditure of products that minimizes packaging (%)
	Best practices 60%	0—1,25—2,5—3,75—5	Number of best practices over the total set related to food waste management

journey towards CE. Additionally, Table 10.2 includes the maximum and minimum values in the scale and the threshold for the 4 evaluation levels that we have established, low, medium, high, very high, level of circularity. Then, each hotel will get an evaluation in each dimension and subdimension and an overall evaluation of circularity based on the points scored in each area.

Conclusions

There is an increasing concern on the part of the hoteliers regarding the sustainability and the waste generated by their respective hotels. Hotels have monitored and measured sustainability indicators so far for a long time. As indicated previously, this approach has been mainly oriented to the reduction of the consumption in energy, water or waste, among others. However, reduction is just compatible with both linear and circular strategies and scraps only on the top of the surface on what it can be achieved. Indeed, there has been great effort on improving areas such as energy efficiency but decisions in a linear model are taken in many cases based purely on power consumption ignoring other factors that are also important for the environment. For instance, the energy and resources consumption for the production of the new devices, the recyclability of the materials of the new devices, the consumption of energy to recycle or dispose the old devices or how old devices are going to be disposed, recycled, etc. In other words, there are far more elements to take into account if we want to be circular than just efficient in terms of energy use. On the other side, sustainability measures have been focusing mainly on the operational part of the hotel business, while CE promotes that circularity starts in the design part. That means, where do we locate the hotel, how do I construct it, which materials do I use, how it is distributed, oriented, etc. There are many decisions to take in this phase that will condition the circularity of the hotel in the long, medium and short term. The problem in this area comes from the scarce circular knowledge, best practice solutions or suppliers that can guide the hoteliers to design more circular hotels.

CE is a complex concept, specially when we think about measuring how circular something is or how one solution might be more circular than another. Therefore, we need to seize it taking small steps to a more circular operation. Additionally, the complexity of Circular Economy and the limited resources that many small hotels have, requires measures that are easy to understand and calculate and that don't disincentive hoteliers in the expected long journey towards circularity. In this chapter we proposed a self-assessment tool to introduce hotels in the evaluation of their degree of circularity.

Currently, there are instruments that have been introduced recently that can gauge an organization's level of circularity and most of the have not yet adopted by the organizations. The complexity of the CE concept and the diversity of the activities that can be in the organizations makes it difficult for these instruments to really fit to the needs of certain industries. In many cases, the instruments are focused on the design of products, they use perception scales, or indicators that cannot be used to compare organizations with different characteristics, such as the size or the level of activity. The importance of sustainability and tourism as a priority industry to meet the environmental goals of the European Union, and the lack of suitable tools addressing industry specifics, suggests that we need to advance in tailored made tools and solutions that can push sustainability actions in hotels. Pioneer initiatives in policy making, such as the Balearic Islands law on urgent measures for the sustainability and circularity of tourism in the Balearic Islands have initiated a path determining priority areas for hotels to move towards CE. It sets an obligation to measure and plan actions through the use of Circularity Plans. Although this is a good starting point, it lacks on indicators that are interesting to evaluate the circular progress and to benchmark hotels circular performance. Additionally, the law does not propose an evaluation scheme which is a key point nowadays to show and market the level of achievement in a certain topic.

In this chapter, we have built on the law's proposal. We created an additional dimension or priority area. We added additional indicators that have been adapted considering industry references for hotel activity, such as room-nights or guest-nights. We included a set of best practices indicators and best practices questionnaire to evaluate the level of sustainable actions of the hotels and as an information tool for hotels where to determine the actions that can be included in their circularity plan. Finally, we created an evaluation scheme which evaluates hotel in a scale 0 to 100 in the 5 dimensions created.

This self-assessment tool is an initial approach and needs a validation step. We have proposed as future lines of research two stages to validate the tool. First, we will submit the tool to a group of experts to determine the usability and potential user experience of the tool, determine additional information or training that might be needed for hoteliers to complete the questionnaire. Second, we will test the tool on a group of hotels to adjust the scales and the evaluation scheme, that is, adjusting the threshold levels, maximum or minimum values, and allocation of point to each of the dimensions.

As indicated in this chapter, the tool has been designed as a starting point to measure, evaluate and manage CE in a hotel. The industry will evolve and circular economy practices will be assumed for more and more hotels, and more indicators and best practices will be required. Thus, we propose a tool that can be scalable, creating one or two more levels in the future. These levels we think that should include more detailed quantitative indicators (i.e. % of textiles that come from organic or recycled materials) more detailed best practices and should be more focused on the quantitative results rather than on the actions taken.

References

Alonso-Almeida, M. M. (2012). Water and waste management in the Moroccan tourism industry: The case of three women entrepreneurs. *Women's Studies International Forum, 35*, 343–353.

Alonso-Almeida, M. M., Robin, C. F., Pedroche, M. S. C., & Astorga, P. S. (2017). Revisiting green practices in the hotel industry: A comparison between mature and emerging destinations. *Journal of Cleaner Production, 140*, 1415–1428.

Alonso-Almeida, M. D. M., Rocafort, A., & Borrajo, F. (2016). Shedding light on eco-innovation in tourism: A critical analysis. *Sustainability, 8*(12), 1262.

Alvarez-Gil, M. J., Burgos-Jimenez, J., & Cespedes-Lorente, J. J. (2001). An analysis of environmental management, organizational context and performance of Spanish hotels. *Omega, 29*, 457–471.

CEEI. (2020). *Self-diagnosis measuring sustainability in organisations*. CEEI Valencia Available from https://ceeivalencia.emprenemjunts.es/?op=65&n=883

CIRCelligence. (2020). *Boston consulting group (BCG)*. Available from https://www.bcg.com/capabilities/social-impact-sustainability/circular-economy-circelligence

Circular IQ. (2020). *Circular transition indicators tool*. Available from https://ctitool.com/

CircularTRANS. (2020). *Mondragón university*. Available from https://www.mondragon.edu/circulartrans/es/login

Circulytics. (2020). *Measuring circularity- ellen MacArthur foundation*. Available from https://www.ellenmacarthurfoundation.org/resources/apply/circulytics-measuring-circularity

COM 102. (2020). *A New Industrial Strategy for Europe*. Communication from the Commission to the European Parliament, the Council, the European Economic and Social Committee and the Committee of the Regions. European Commission.

COM 640. (2019). *The European Green Deal*. Communication from the Commission to the European Parliament. the European Council, the Council, the European Economic and Social Committee and the Committee of the Regions.

COM 98. (2020). *A New Circular Economy Action Plan for a Cleaner and More Competitive Europe*. Communication from the Commission to the European Parliament, the European Council, the Council, the European Economic and Social Committee and the Committee of the Regions.

Comunidad Autónoma de las Illes Balears. (2022). *Ley 3/2022, de 15 de junio, de medidas urgentes para la sostenibilidad y la circularidad del turismo de las Illes Balears*. Available from https://www.boe.es/buscar/doc.php?id=BOE-A-2022-13846

Confederación Española de Hoteles y Alojamientos Turísticos CEHAT. (2008). *Manual de buenas prácticas para la mejora de la eficiencia energética de los hoteles de Canarias*.

Fernández-Robin, C., Celemín-Pedroche, M. S., Santander-Astorga, P., & Alonso-Almeida, M. D. M. (2019). Green practices in hospitality: A contingency approach. *Sustainability, 11*, 3737.

Florido, C., Jacob, M., & Payeras, M. (2019). How to carry out the transition towards a more circular tourist activity in the hotel sector. *The role of innovation. Administrative Sciences, 9*, 47.

Franco, N. G., Almeida, M. F. L., & Calili, R. F. (2021). A strategic measurement framework to monitor and evaluate circularity performance in organizations from a transition perspective. *Sustainable Production and Consumption, 27*, 1165–1182.

Ghisellini, P., Cialani, C., & Ulgiati, S. (2015). A review on circular economy: The expected transition to a balanced interplay of environmental and economic systems. *Journal of Cleaner Production, 14*, 11–32.

Girard, L. F., & Nocca, F. (2017). From linear to circular tourism. *Aestimum, 70*, 51–74. https://www.wttc.org/about/media-centre/press-releases/press-releases/2019/travel-tourism-continuesstrong-growth-above-global-gdp/ (accessed on 19 August 2019)

Hamele, H., & Eckardt, S. (2006). *Environmental initiatives by European tourism businesses: Instruments, indicators and practical examples*. ECOTRANS, IER.

InnoEcoTur. (2022). *Informe de necesidades del sector turístico para la transición a la economía circular*. Available from https://innoecotur.webs.upv.es/primer-informe-de-necesidades-del-sector-turistico-para-la-transicion-a-la-economia-circular/

InnoEcoTur. (2023). *Transición del sector turístico a la economía circular. Necesidades, retos y mejoras*. Available from https://innoecotur.webs.upv.es/transicion-del-sector-turistico-a-la-economia-circular-necesidades-retos-y-mejoras/

Kirchherr, J., Reike, D., & Hekkert, M. (2017). Conceptualizing the circular economy: An analysis of 114 definitions. *Resources, Conservation and Recycling, 127*, 221–232.

Kravchenko, M., Pigosso, D. C., & McAloone, T. C. (2019). Towards the ex-ante sustainability screening of circular economy initiatives in manufacturing companies: Consolidation of leading sustainability-related performance indicators. *Journal of Cleaner Production, 241*, 118318.

Kristensen, H. S., & Mosgaard, M. A. (2020). A review of micro level indicators for a circular economy–moving away from the three dimensions of sustainability? *Journal of Cleaner Production, 243*, 118531.

Lenzen, M., Sun, Y., Faturay, F., Ting, Y., Geschke, A., & Malik, A. (2018). *The carbon footprint of global tourism. Nature climate change.* Macmillan Publishers Limited.

Lindgreen, E. R., Salomone, R., & Reyes, T. (2020). A critical review of academic approaches, methods and tools to assess circular economy at the micro level. *Sustainability, 12*(12), 4973.

Manniche, J., Topsø Larsen, K., Brandt Broegaard, R., & Holland, E. (2017). Destination: A circular tourism economy: a handbook for transitioning toward a circular economy within the tourism and hospitality sectors in the South Baltic Region; Project Mac-CIRTOINNO; Centre for Regional & Tourism Research (CRT): Nexø, Denmark.

Marino, A., & Pariso, P. (2021). The transition towards to the circular economy: European SMEs' trajectories. *Entrepreneurship and Sustainability Issues, 8*(4), 431.

Moraga, G., Huysveld, S., Mathieux, F., Blengini, G. A., Alaerts, L., Van Acker, K., et al. (2019). Circular economy indicators: What do they measure? *Resources, Conservation and Recycling, 146*, 452–461.

Parchomenko, A., Nelen, D., Gillabel, J., & Rechberger, H. (2019). Measuring the circular economy – A Multiple Correspondence Analysis of 63 metrics. *Journal of Cleaner Production, 210*, 200–216.

PriceWaterhouseCoopers. (2013). *Cómo impulsar la eficiencia energética en el sector hotelero español.* Available from https://www.pwc.es/es/publicaciones/energia/assets/como-impulsar-la-eficiencia-energetica-en-el-sector-hotelero-espanol.pdf

Ricaurte, E., & Jagarajan, R. (2021). *Hotel Sustainability Benchmarking Index 2021: Carbon, energy, and water.*

Rodríguez, C., Florido, C., & Jacob, M. (2020). Circular economy contributions to the tourism sector: A critical literature review. *Sustainability, 12*(11), 4338.

Saidani, M., Yannou, B., Leroy, Y., & Cluzel, F. (2017). How to assess product performance in the circular economy? Proposed requirements for the design of a circularity measurement framework. *Recycling, 2*(1), 6.

Seguí, L., Medina, R., & Guerrero, H. (2018). Gestión de residuos y economía circular. *EAE Business School*, 1–46.

Singh, P., & Giacosa, E. (2019). Cognitive biases of consumers as barriers in transition towards circular economy. *Management Decision, 57*, 921–936.

Sorin, F., & Einarsson, S. (2020). *Circular economy in travel and tourism: a conceptual framework for a sustainable, resilient and future proof industry transition.*

Styles, D., Schoenberger, H., & Galvez, M. J. (2013). *Best environmental management practice in the tourism sector.* Publications Office of the European Union.

United Nation World Tourism Organization UNTWO. (2008). *United Nations Environment Programme UNEP. Climate Change and Tourism.* Responding to Global Challenges; World Tourism Organization: Madrid, Spain.

United Nation World Tourism Organization UNTWO. (2022). *UNWTO World Tourism Barometer*, Vol. 20, Issue 5, September. Available from https://webunwto.s3.eu-west-1.amazonaws.com/s3fs-public/2022-09/UNWTO_Barom22_05_Sept_EXCERPT.pdf?VersionId=pYFmf7WMvpcfjUDuhNzbQ_G.4phQX79q

United Nations-UN United Nations Treaty Collection Chapter XXVII Environment, Paris Agreement, Paris, 12 December 2015. *New York, United Nations* 2016. Available online: https://treaties.un.org/pages/Treaties.aspx?id=27&subid=A&clang=_en (accessed on 19 August 2019).

Vinante, C., Sacco, P., Orzes, G., & Borgianni, Y. (2021). Circular economy metrics: Literature review and company-level classification framework. *Journal of Cleaner Production, 288*, 125090.

Vourdoubas, J. (2016). Energy consumption and use of renewable energy sources in hotels: A case study in Crete, Greece. *Journal of Tourism and Hospitality Management, 4*(2), 75–87.

World Travel & Tourism Council (WTTC). Annual Report. (2019). Available from https://www.wttc.org/about/media-centre/press-releases/press-releases/2019/travel-tourism-continuesstrong-growth-above-global-gdp/

Open Access This chapter is licensed under the terms of the Creative Commons Attribution 4.0 International License (http://creativecommons.org/licenses/by/4.0/), which permits use, sharing, adaptation, distribution and reproduction in any medium or format, as long as you give appropriate credit to the original author(s) and the source, provide a link to the Creative Commons license and indicate if changes were made.

The images or other third party material in this chapter are included in the chapter's Creative Commons license, unless indicated otherwise in a credit line to the material. If material is not included in the chapter's Creative Commons license and your intended use is not permitted by statutory regulation or exceeds the permitted use, you will need to obtain permission directly from the copyright holder.

Chapter 11
Conclusions: Tourism Sustainability and Improvement Plans

Ángel Peiro-Signes and Virginia Santamarina-Campos

As this book concludes, it is essential to summarise the critical mission we have engaged in: facilitating a shift towards the circular economy within the expanding tourism industry in the Valencian Region. This complex task has involved a multidimensional approach combining rigorous academic research, empirical field analyses, active collaboration with key stakeholders, and the development of the InnoEcoTur Innovation Platform.

This conclusion brings together the collective wisdom and insights gained, extracting valuable viewpoints and emerging ideas from our core sections: 'Challenges and Opportunities,' 'Best Practices', and 'Research, Innovation, Competitiveness, and Production'. The closing remarks not only signify the culmination of our efforts but also serve as a comprehensive summary that outlines the practical steps for advancing towards a more sustainable, circular economy in the region's tourism industry.

Insights for Action

Our study of the challenges and opportunities has revealed that a significant shift in 'impact culture' is essential for a successful transition to a circular economy model. Rather than viewing sustainability as a burdensome obligation, the industry needs to

---

Á. Peiro-Signes
Department of Business Organisation, Universitat Politècnica de València, Valencia, Spain
e-mail: anpeisig@omp.upv.es

V. Santamarina-Campos (✉)
Department of Conservation and Restoration of Cultural Assets, Universitat Politècnica de València, Valencia, Spain
e-mail: virsanca@upv.es

© The Author(s) 2024
M. Segarra-Oña et al. (eds.), *Managing the Transition to a Circular Economy*, SpringerBriefs in Business, https://doi.org/10.1007/978-3-031-49689-9_11

perceive it as a strategic asset that can offer both economic and environmental dividends.

Our 'Good Practices' section is more than just a list of successful case studies; it is a template for action. The varied applications of circularity in hotels, restaurants, and even specialised tourism segments like wine and beer tourism, underscored that circular practices are versatile enough to be tailored to different business models and scales.

Finally, our in-depth look into 'Research, Innovation, Competitiveness and Production' revealed that the circular economy isn't just an environmental strategy; it's a comprehensive business model that can elevate an organisation's competitive advantage. The relationship between green innovation and corporate performance signalled that businesses which are progressive in their environmental strategies often excel in their market performance as well.

Policy Implications and Future Work

This book was not conceived as an endpoint, but rather as a dynamic tool that can evolve. Building on the foundation laid by InnoEcoTur, the work ahead involves translating these insights into policy initiatives and tangible business practices. Close collaboration between universities, government agencies, and the tourism industry will be vital in making this translation a success to deliver major impact.

The Valencian Innovation Agency, together with the Valencian universities and other political and business actors, can play a driving role in mainstreaming these best practices through policy frameworks. Similarly, future iterations of this project could delve deeper into the application of circular economy principles in emerging tourism sub-sectors and evaluate their long-term impacts.

A Call to Action

We initiated this project with a sense of urgency, aware of the environmental impacts overshadowing the success of the tourism industry. As the project draws to a close, this urgency has been transformed into a sense of possibility. The circular economy isn't a mere theoretical construct; it's a roadmap for a sustainable future in Valencian tourism. We hope this book serves as both a guide and a catalyst, inspiring businesses and policymakers alike to enact change that is both profitable and sustainable.

In conclusion, the InnoEcoTur project has initiated a pivotal dialogue, and now it falls to all of us—academics, policymakers, and industry practitioners—to continue to move forward and turn it into actionable change. Here's to charting a more circular, sustainable, and prosperous course for the Valencian Region's tourism industry.

Open Access This chapter is licensed under the terms of the Creative Commons Attribution 4.0 International License (http://creativecommons.org/licenses/by/4.0/), which permits use, sharing, adaptation, distribution and reproduction in any medium or format, as long as you give appropriate credit to the original author(s) and the source, provide a link to the Creative Commons license and indicate if changes were made.

The images or other third party material in this chapter are included in the chapter's Creative Commons license, unless indicated otherwise in a credit line to the material. If material is not included in the chapter's Creative Commons license and your intended use is not permitted by statutory regulation or exceeds the permitted use, you will need to obtain permission directly from the copyright holder.

MIX
Papier aus verantwortungsvollen Quellen
Paper from responsible sources
FSC® C105338

If you have any concerns about our products,
you can contact us on
ProductSafety@springernature.com

In case Publisher is established outside the EU,
the EU authorized representative is:
**Springer Nature Customer Service Center GmbH
Europaplatz 3, 69115 Heidelberg, Germany**

Printed by Libri Plureos GmbH
in Hamburg, Germany